ENCYCLOPAEDIA OF ENVIRONMENTAL BIOTECHNOLOGY

ENCYCLOPAEDIA OF ENVIRONMENTAL BIOTECHNOLOGY

Vol. 2

Shahzad Ahmad

ANMOL PUBLICATIONS PVT. LTD.
NEW DELHI - 110 002 (INDIA)

ANMOL PUBLICATIONS PVT. LTD.
H.O.: 4374/4B, Ansari Road, Darya Ganj,
New Delhi-110 002 (India)
Ph.: 23278000, 23261597
B.O.: No. 1015, Ist Main Road, BSK IIIrd Stage
IIIrd Phase, IIIrd Block,
Bangalore - 560 085 (India)
Visit us at: www.anmolpublications.com

Encyclopaedia of Environmental Biotechnology

First Published, 2008

ISBN 978-81-261-3539-4 (Set)

PRINTED IN INDIA

Printed at Mehra Offset Press, Delhi.

Contents

Preface

Environmental biotechnology is the applied technology for the study of natural environment. Environmental biotechnology could also imply that one tries to harness biological process for commercial use and exploitation. Environmental biotechnology, by contrast, deals readily and apparently with dramatic topics, and though their importance, albeit different, may be every bit as great as their direct relevance is quite readily appreciated by the bulk of population. Cleaning up contamination and dealing rationally with waste is of course, in everybody's best interests, but for most people, this is like simply addressing a problem. Some of the potentially beneficial uses of biological engineering, which touch the lives of the majority of people, however directly involve much simpler approaches. In many respects, there is a logical link between the aims of environmental biotechnology.

Environmental biotechnology is fundamentally rooted in wastes, in its various guises, typically being concerned with the remediation of the contamination caused by previous use and the impact reduction of current activity or the control of pollution. Thus, the principal aim of this stream are the manufacture of products in environmentally harmonious manner, which allow for the minimisation of harmful solids, liquids or gaseous outputs or clean up the residual effects of earlier human occupations. Environmental biotechnologists may enhance or optimise conditions for existing biological systems to make their activities happen faster or more efficiently or they resort to some form of alteration to bring about the desired results. The variety of organism, which may play a part in environmental applications of biotechnology is wide and ranging from microbes through trees and all are utilised on one of the same three fundamental bases.

Accordingly, the range of businesses to which environmental biotechnology has potential relevance is almost limitless. All commercial operations generate wastes in one form or the other. Additionally, biotechnology can be of further commercial significance by converting low-cost organic feed stocks into high-value products.

Biotechnology based methods are seen as essential to help meet European Union (EU) targets for biowaste diversion and reductions in pollutants. Across the world, the existing regulations on environmental pollution are required to be rigorously enforced, with more stringent compliance standards implemented. All of this is expected to stimulate the sale of biotechnology based environmental processing methods significantly and in particular, the global market share is projected to grow faster than the general biotic sector trends.

This exhaustive work has a perfect treatment with environmental biotechnology in an objective manner. Expectedly, this work would be capable to create a sort of awareness among the people and make the study of environmental biotechnology easier and more meaningful.

Editor

1

Role of Biosensors

In the 1980s there have been rapid advancement in the application of molecular biology and other biological techniques to create new processes and products of commercial interest which we now call biotechnology. While genetic engineering and monoclonal antibody technology are prominent leaders at present, the coupling of immobilised enzymes and other biological materials to microelectronics promises to be another fertile area in biotechnology. It is from the wedding of biochemistry and microelectronics that biosensors have made appearance. When fully developed they will form the basis of many analytical systems with such varied applications as health care and industrial monitoring. In the future it is hoped that the information gained from the construction of miniaturised biosensors will lead to the discovery of workable molecular switching devices—true biochips—which can be used for computer construction.

Concept of Biosensor

It is a device made by fusion of a biocomponent and an electronic gadget, used to detect the presence of the substance to be monitored. It works on the principle that when the biocomponent of the biosensor comes in contact with the substance (to be monitored) a reaction takes place, which triggers the electronic gadget to produce a signal, indicating the presence or absence of the substance. Thus, the biosensor can be defined as an analytical device or system consisting of an immobilised biological material (such as an enzyme, antibody, organelle or a whole cell) in intimate contact with a suitable transducer device. The latter converts the biochemical signal into a quantifiable electrical signal (Gronow *et al.*, 1988).

A biosensor combines the selectivity of biology with the processing power of modern microelectronics and optoelectronics to offer powerful new analytical tool with major application in median, environmental diagnostics and the food and processing industries.

The conventional aim of a biosensor is to produce either discrete or continuous digital electronic signals which are proportional to a signal analyte or a related group of analytes.

Immobilisation

A crucial step in the development of a biosensor is immobilisation of the biocomponent without adversely affecting its activity and stability. There are different methods by which biomolecules are immobilised, these are:

Adsorption

This is the oldest technique. The biological material is simply contacted with a surface-active material such as graphite, alumina, clay, glass, or iron-exchange resin. A major disadvantage is that desorption frequently occurs following variation in pH; ironic strength, temperature, solvent, etc. However, there have been notable successes with antibodies and certain enzymes in achieving stable adsorption.

Covalent Attachment

Chemical groups on the biological material not essential for activity (e.g., non-essential amino acid residues of enzymes) are attached to chemically activated supports such as synthetic polymers, cellulose, glass, etc. The linking molecule may form a bridge (e.g., when using cyanuric chloride) or be expelled after coupling (e.g., carbodi-imides).

Cross-linking

Intermolecular covalent linkages can be formed between macromolecules with the aid of bifunctional cross-linking agents (e.g. di-isothiocynates, alkylating agents, and dialdehydes such as glutaraldehyde). A major disadvantage is that some systems lose biological activity when their three-dimensional structure is altered in this way. This may be minimised by the inclusion of inert carrier molecules such as serum albumin.

Physical Entrapment

Gels or polymers such as polyacrylamides, silica gel, and starch are cross-linked in the presence of the enzyme or other biological material which is thereby trapped or captured. The pores of the matrix are large enough to allow substrate and product diffusion but small enough to prevent loss of the active high molecular weight biological material. One method of particular relevance to biosensors is to entrap within the pores of asymmetric ultrafiltration membranes, giving thin, highly active layers of biomaterial.

Microencapsulation

Materials are trapped by the membranes of various polymers, usually in capsules with mean diameters ranging from 5 to 300 μm. The entrapped biological material is too large to mote through the pores of the semi-permeable synthetic membrane, but smaller substrate and product molecules can move readily through the pores of the membrane.

With the exception of microencapsulation these methods have all been of importance in biosensor configuration. Membranes in particular have been used as the immobilisation support as they allow a think layer of active material to be positioned close to the surface of a transducer. In this way enzyme electrodes in particular can be constructed with fast, linear responses. Whole cells and organelles have also received considerable attention. It should be noted that the preparation of highly active and stable materials for use in large-scale biotechnology is the main the groundwork for the successful use of immobilised biological material in analytical devices.

Construction of Biosensor Devices

Biosensor consists of two primary components—(a) biological, e.g. enzyme antibody and (b) physical, e.g., transducer, amplifier, etc. The biological analyte and the physical analyte interact in such a manner which produces some physical charge detectable by the transducer. These properties of the biological component impart on the biosensor its specificity, sensitivity and the ability to detect and measure the analyte. The biological component is suitably immobilised on to the transducer (Enzymes are usually immobilised by glutaraldehyde on a porous sheet like lens tissue paper or nylon net fabric; the enzyme membrane so produce is affixed to the transducer). Generally correct immobilisation of enzymes enhances their stability, which may be rather dramatic in some cases. As a result, any enzyme immobilised systems can be more than 10000 times over a period of several months.

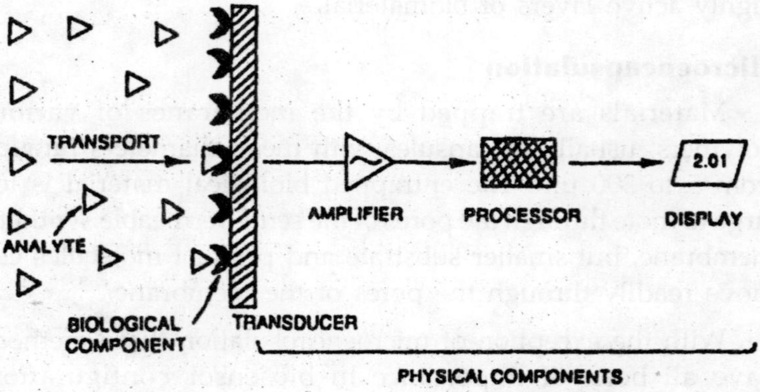

A schematic representation of the various components of a biosensor. The biological component may be enzyme, nucleic acid, antibody, etc. The analyte must be transported from the solution to the biological component for the reaction, ordinarily the transport is due to simple diffusion.

The biological component interacts specifically to the analyte which produces a physical change close to the transducer surface. This physical change may be: (i) heat released or absorbed by the reaction (measured by calorimetric biosensor), (ii) production of an electrical potential due to changed distribution of electron (potentiometric biosensors), (iii) movement of electrons due to redox reaction (amperometric biosensors), (iv) light produced or absorbed during the reaction (optical biosensors), or (v) change in mass of the biological component as a result of the reaction (acoustic wave biosensors).

The transducer detects and measures this change and converts it into an electrical signal. This signal is necessarily very small, and is amplified by an amplifier before it is feed into the microprocessor. The signal is than processed and interpreted, and is displayed in suitable units. Thus biosensors convert a chemical information flow into an electrical information flow involving the following steps.

1. The analyte diffuses from the solution to the surface of the biosensor.
2. The analyte reacts specifically and efficiently with the biological component of the biosensor.
3. This reaction changes the physicochemical properties of the transducer surface.
4. This leads to a change in the optical or electronic properties of the transducer surface.
5. The change in optical / electronic properties is measured, converted into electrical signal which is amplified, processed and displayed.

An efficient biosensor must have at least some of the following features:

1. It should be highly specific for the analyte.

2. The reaction used should be independent of such factors as stirring, pH, temperature, etc.
3. The response should be linear over a useful range of analyte concentrations.
4. The device should be small and biocompatible, if it is to be used for analyses within the body.
5. The device should be cheap, small and easy to handle and use.
6. It should be durable, i.e., capable of repeated use.

A list of some biological materials and optically or electronically active devices commonly used in biosensors.

Biological material	*Detection devices (or transducer)*	*Examples*
Enzymes	1. Potentiometric electrodes	1. Enzyme electrode for urea (based on urease)
Nucleic acids	Amperometric electrodes	2. Enzyme electrode for organophosphorus pesticides (using acetyl cholinesterase).
Antibodies	Wave guides (optical biosensor)	
Lictins	Grating coupler (optical biosensor)	
Cells	Acoustic wave sensors	
Organs	Conductimetric sensors	
Tissue slices	Thermometric sensors	

Types of Biosensors

The biosensors are of mainly of five types: (1) Calorimetric, (2) Potentiometric, (3) Amperometric, (4) Optical and (5) Acoustic wave biosensor. The main features of these biosensors are briefly described below.

Calorimetric Biosensors

Many enzyme catalysed reactions produce heat (exothermic). Calorimetric biosensors measure the change in temperature of the solution containing the analyte following enzyme action and interpret it in terms of the analyte concentration in the solution. The analyte solution is passed through a small (ca. 1ml) packed bed column containing immobilised enzyme; the temperature of the solution is determined just before entry of the solution into the column and just as it is leaving the column using separate thermistors.

This is the most generally applicable type of biosensor, and it can be used for turbid and strongly coloured solutions. The greatest disadvantage is to maintain the temperature of the sample stream *at* a constant, say ± 0.01°C, temperature. The sensitivity (10^{-4} M) and the range (10^{-4} –10^{-2} M) of such biosensors is quite low for most applications. The sensitivity can be increased by using two or more enzymes of the pathway in the biosensor to link several reactions to increase the heat output. Alternatively, multifunctional enzymes may be used. An example is the use of glucose oxidase for determination of glucose.

Potentiometric Biosensors

These biosensors use ion-selective electordes that convert the biological reaction into electronic signal. The electrodes employed are most commonly pH metre glass pH electrodes (for cations), glass elctrodes coated with a gas selective membrane (for CO_2, NH_3 or H_2S) or solid state electrodes. Many reactions generate or use up H^+ which is detected and measured by the biosensor, in such cases very weakly buffered solutions are used. Gas sensing electrodes detect and measure the amount of gas produced. An example of such an electrode is based on urease which catalyses the following reaction.

$$CO(NH_2)_2 + 2H_2O + H^+ \rightarrow 2NH^+_4 + HCO_3$$

The reaction can be measured by a pH sensitive, ammonium ion sensitive, NH_3 sensitive or CO_2 sensitive electrodes. Biosensors can now be prepared by placing coated membranes on the ion-selective gates of ion-sensitive field effect transistor. These biosensors are extremely small (« 0.1 mm²).

Amperometric Biosensors

These biosensors function by the production of current when potential is applied between two electrodes, the magnitude of current being proportional to the substrate concentration. The simplest amperometric biosensors use the Clark oxygen electrode which determines the reduction of oxygen present in the sample (analyte) solution (These are the first generation biosensors). These biosensors are used to measure redox reactions, a typical example being the determination of glucose using glucose oxidase.

The major problem of such biosensors is their dependence on the dissolved oxygen concentration in the analyte solution. This may be overcome by using mediatros, e.g., ferrocenes; these molecules transfer the electrons generated by the reaction directly to the electrode rather than reducing the oxygen dissolved in analyte solution (these are called second generation biosensors). The combination of a stable enzyme, the use of a mediator to reduce the operating potential (thus eliminating the need for anti-interference membranes), and the adequacy of reference electrodes that are not ideal, have enabled MediSence Inc. (Cambridge, MA, USA) to mass-produce successfully a disposable senor for measuring blood-glucose levels by diabetic patients. The present-day electrodes, however, remove the electrons directly from the reduced enzymes (without the help of mediators), and are coated with electrically conducting organic salts.

Optical Biosensors

These biosensors measure both catalytic and affinity reactions. They measure a change in the fluorescence or in absorbance caused by the products of catalytic reaction. Alternatively, they measure the changes induced in the intrinsic optical properties if the biosensor surface due to loading on it of dielectric molecules like protein (in case of affinity reactions). A most promising biosensor involving luminescence use firefly enzyme luciferase for detection of bacteria in food or clinical samples. The bacteria are specifically lysed to release ATP, which is used by luciferase in the presence of oxygen to produce light which is measured by the sensor.

Acoustic Wave Biosensor

These are also called piezoelectric devices. Their surface is usually coated with antibodies which bind to the complementary antigen present in the sample solution. This leads to increased mass which reduces their vibrational frequency. This change is used to determine the amount of antigen present in the sample solution.

Commercial Biosensors

Biosensors represent a rapidly expanding area of technology promising economic activity and growth. The estimated world market is approximately $25 billion per year of which 30 per cent is towards health care. At present only 0.1 per cent of this market is using biosensors. Biosensors market is likely to expand rapidly with the development of low cost, easy to use and durable biosensors.

To be commercially successful, a biosensor ought to meet the general requirements fulfilled by commercial sensors. The biosensors must have:

1. Relevance of output signal to measurement environment,
2. Accuracy and repeatability,
3. Sensitivity and resolution,
4. Dynamic range,
5. Sped of response,
6. Insensitivity to temperature (or temperature compensation),
7. Insensitivity to electrical and other environmental interferences,
8. Amenability to testing and calibration,
9. Reliability and self-checking capability,

10. Servicing facilities,
11. Bearable running costs,
12. Acceptability by user and
13. Product safety host system must not be contaminated by sensor.

The model as illustrated in the following shows the characteristics of a wide range of biosensor devices which have been demonstrated in the scientific literature, or as commercial products.

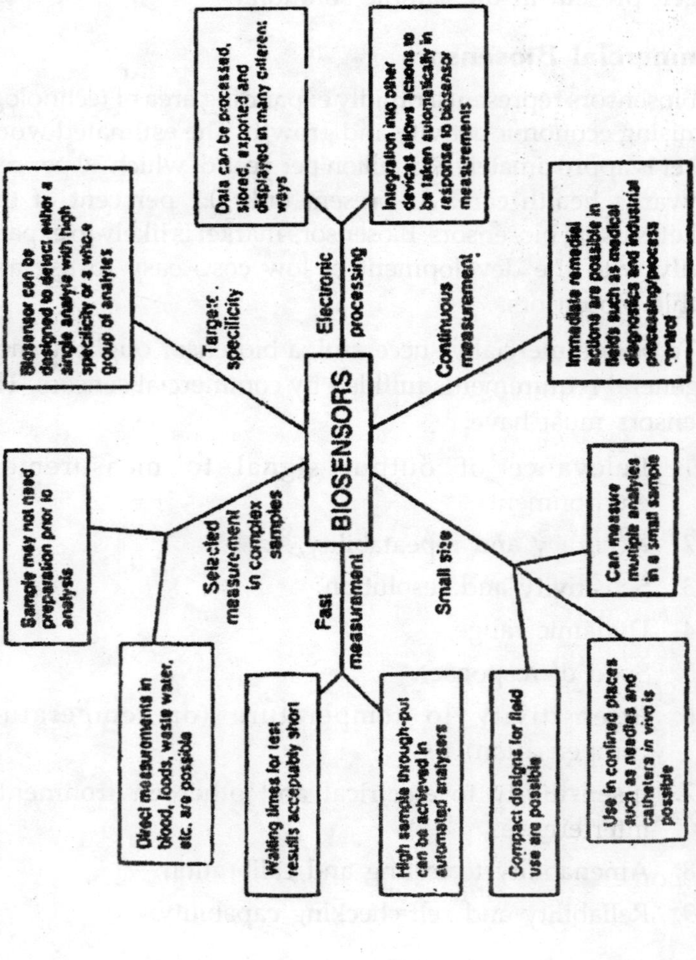

Characteristics of Ideal Biosensor

Biosensor Development: The major tasks and skills in developing a biosensor for a target analyte include:

1. Selection of a suitable bioreceptor molecule. This requires knowledge in biochemistry and biology.
2. Selection of a suitable immobilisation method. This requires knowledge in chemistry.
3. Selection of a suitable transducer. Knowledge in electrochemistry and physics is necessary.
4. Designing a biosensor considering measurement ranger, linearity, and minimisation of interference. Knowledge in kinetics and mass transfer are required for this purpose.
5. Packaging of biosensor—Once a biosensor has been designed, it has to be put into a package for convenience in manufacturing and use. The current trend is miniaturisation and mass production. Modern IC (integrated circuit) fabrication technology and micromachining technology are used increasingly in fabricating biosensors.

It is clear that interdisciplinary cooperation is essential for a successful development of a biosensor.

Biosensor Diversification

Most recent work on electrochemical methods has concentrated on two main areas, namely redox-mediated amperometric sensors and field effect transistor (FET)-based devices.

Sensors based on measurement of heat (e.g., thermistor devices) and conductance have also received attention. Additionally, there has been considerable interest in optoelectronic methods based on fibre-optics and such phenomena as surface Plasmon resonance and evanescent waves.

Redox-mediated Systems

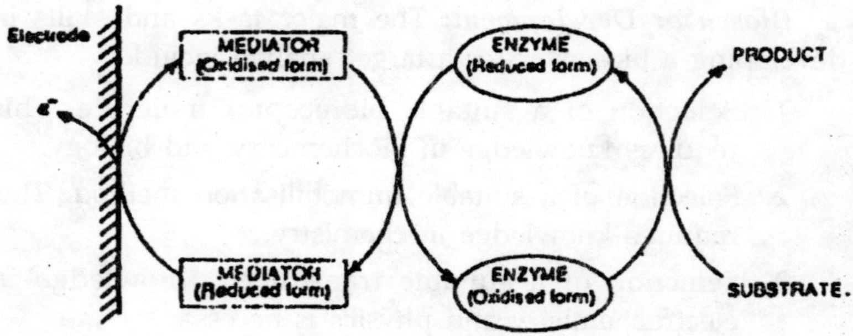

Working principle of mediated biosensors

The redox-mediated enzyme electrode arose from the work carried out on biofuel cells. The main aim in manufacturing of this type of biosensor is to facilitate of electrons generated by an oxidoreductase enzyme (or enzyme system) to the electrode surface.

Natural mediators such as the cytochromes have been shown to promote passage of electrons but redox-active dyes and other artificial mediator compounds have proved more useful: one of the most extensively studied of these are ferrocene and its derivatives. In the case of an oxidase the mediator is able to replace the requirement of the enzyme for oxygen, thus eliminating dependence on the oxygen content of the sample. Mediators may also be used with certain NAD-independent dehydrogenase such as lactate dehydrogenase (cytochrome b2) and the quinoproteins. By varying the structure of the compound used it is possible to tailor a mediator to meet specific requirements, for instance giving low solubility to aid retention at the electrode surface, or changing the half-wave potential of the mediator to reduce the required operating potential and minimise interference. This sort of flexibility has made use of mediators and matter of choice in many configurations. However, mediators also have some disadvantages, perhaps the most important of which is the lack of reproducibility for reagentless biosensors, i.e., those in which mediator is not added to the sample. In addition to their use in enzyme electrodes, mediators are suitable for immunosensing applications.

If oxygen and peroxide-based system are classified as 'first generation' enzyme electrodes, and mediated systems are the second generation, then the third generation is that which uses direct electron transfer (DET). The oxidation and reduction of glucose oxidase have been shown on certain forms of carbon, although this is usually unsuitable for biosensors owing to the low signals so far achieved. Conducing organic salts resulting from use of tetracyanoquinodimethane (TCNQ) have been claimed to promote DET. Although this approach has advantage over mediated systems for being entirely reagentless, in practice it is likely to be limited to a small number of enzymes.

In addition to the use of soluble mediators, chemically modified electrodes have generated considerable interest. These consist of conventional electrodes, usually of carbon which have been coated with a reagent to assist electron transfer at electrode surface. A number of types have been described which allow reproducible oxidation of NADH, paving the way for a series of enzyme electrodes based on dehydrogenases.

FETs (Field Effect Transistors)

Over the past 10 years strenuous efforts have been made to produce a miniaturised electrochemical type of biosensor using conventional electronic devices called field effect transistors (FETs). When suitably modified, these very small devices have the ability to measure charge accumulation at their surface and they can be modified to be pH-sensitive. FETs can be made to measure ions by using selective materials over the sensitive gate of the FET thus giving ion-selective FETs (ISFETs) similar in use to a miniature ISE. By adding enzymes (enFETs) or antibodies (immunoFETs) a range of true biosensors may be contemplated, very small and cheap enough in construction to be disposable. However, the fundamental manufacturing problems and selectivity of these devices, and the difficulty in miniaturisation of a suitable reference electrode, their practical use has been presently restricted to ion and gas measurements.

A recent interesting variant of this technology has been described by a California based company (Molecular Devices Corp.). Here an LED is used to create a photoresponse in a silica-based chip, this being recorded as an AC current. As the potential across the device is changed by the external power supply the capacitance is changed and thus the magnitude of the AC current. A biochemical reaction at the surface can induce a measurable effect on the current. This is illustrated below.

Novel Photoresponse Silicon Chip Biosensor Device

Although still at an early developmental stage this represents an interesting combination of electrochemically and optically based biosensing techniques.

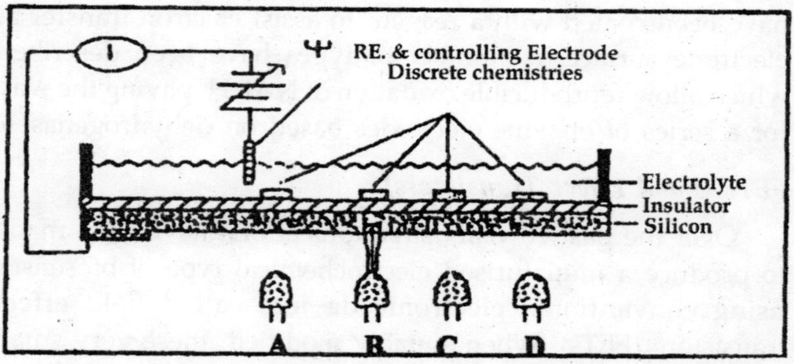

Thermistors

One interesting development from the 1979s onwards involves utilisation of thermistor to monitor the small temperature differences generated in many biochemical reactions. A linear response to temperature changes in the range 0.01-0.001°C can be obtained. Sensors based on the measurement of heat generated from an immobilised enzyme column have been described, although the use of a column upstream from the thermistor means these are not true 'biosensors' in the strict sense. The need for effective thermal screening of such devices makes miniaturisation into a biosensor format very difficult. Some examples of assays performed with the enzyme thermistor are shown below:

Example of assays performed with the enzyme thermistor

Substance	*Enzyme(s)*	*Concentration range* x 10^{-6} *mol* dm^{-3}
Ascorbic acid	Ascorbic acid oxidase	50 - 100
Cholesterol	Cholesterol oxidase	30 - 150
Glucose	Glucose oxidase + catalate	700
Phenol	Tyrosinase	100 - 1000
Sucrose	Invertase	50 - 100000

Conductimeters

The conductimeters initially seemed very promising as a transducer capable of detecting a wide range of enzymatic reactions. Two sets of electrodes are used in a planar configuration with enzyme immobilised across one set; the other set acts as a blank for background conductance. Theoretically, any reaction producing charged species may be detected by this method. In practice, the high and variable background conductivity of biological solutions make this technique too insensitive for most applications.

Piezoelectric Crystals

The piezoelectric crystal has been put forward as a very sensitive transducer for biosensor applications. Binding of a molecular species to the crystal surface causes a mass change that is sensitively monitored as a change in oscillating frequency. Owing to the damping effect of immersion of the crystal into liquids, useful sensitivity can only be achieved for analytes in the gas phase. Organophosphorous compounds have been detected in air by immobilising the enzyme acetylecholinestrease on to a crystal, a device which could be used to detect nerve gas. By using selective membrane coating, gases such as carbon monoxide and ammonia can be measured. By immobilising antibodies or special 'receptor molecules' to these crystals, there arises the possibility of making selective 'electronic noses' for detecting illegal drugs.

Optoelectronic Biosensors

Optoelectronic biosensors were first worked out towards the end of the 1970s which is represented diagrammatically in the following.

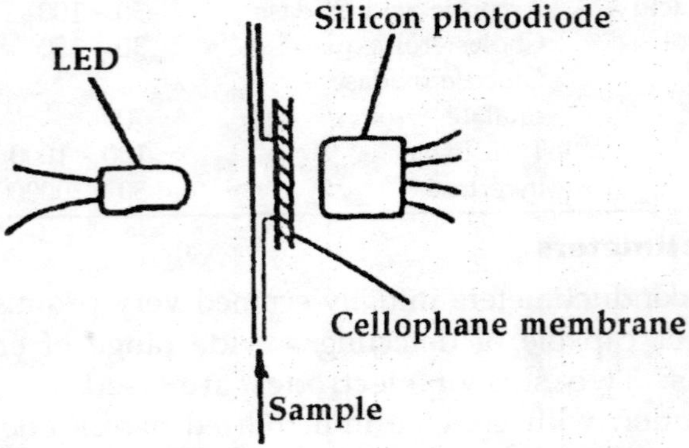

Outline of optoelectronic sensor

In recent years there has been a great deal of interest in optoelectronic biosensors, particularly for their potential sensitivity, the variety of techniques available, and the possibility that optically based assays may be readily converted into a biosensor configuration. Several systems have been described which utilise fibre optics, for instance, a dye indicator has been immobilised at the end of an optical fibre to make a miniature pH sensor. As with FETs, the pH-sensitive transducer can then be used to construct more sophisticated systems using immobilised enzymes or antibodies. In addition to measurement of absorption of light, reflectance, fluorescence, and chemiluminescence may be used, giving the fibre-optic approach a wide versatility.

Practical Biosensors

Practically, four forms of biosensor are emerging for medical and industrial applications.

Small, Hand-held Devices: These are designed monitoring of blood glucose levels in diabetics. The sensor configuration may be as a lipstick pen-shaped device, a disposable enzyme-loaded membrane, or a semi-permanent electrode covered by a permeable membrane. Each of these methods had its own particular merits, but the semi-permanent electrode allows quality control and calibration to be carried out on the actual sensor element to be used unlike a disposable strip. Devices for the measurement of other clinical parameters at home and in the doctor's chamber are also under development. The major requirements for such devices are robustness ease of use by unskilled persons, small size, fast speed of measurement and an easily readable display.

The Laboratory Analyser: The use of biosensors in laboratory clinical chemistry potentially confers several advantages over currently employed methods of analysis. The use of undiluted blood measurement allows:

(a) Faster analysis, since no separation step is required.

(b) Avoidance of errors in pipetting and dilution.

(c) The manufacture of a completely self-contained unit requiring no external resources other than a power supply.

(d) Measurements which approximately more closely to the *in vivo* situation than using serum or plasma.

Currently, clinical laboratory testing is mainly carried out on multi-channel analysers. These are capable of measuring many parameters in parallel or in series, and can quickly analyse large numbers of samples. Adoption of these systems is seen in Technion auto-analysers which are already in use in many laboratories. Many biosensor configurations are constructed in-house, such as a recently made nylon coil device for an automated analyser having a useful lifetime of six months. Complementary to the multi-channel systems are 'star' clinical laboratory anlysers, designed for rapid measurement of single or small numbers of samples. These are usually small, discrete instruments, often transportable between laboratories and clinics. The main demand is for a glucose instrument for diabetic clinics. One of the few commercially successful bench-top analysers is the Yellow Spring YSI 23A glucose analyser which uses immobilised glucose oxidase between two polymer membranes: the instrument incorporates a dilution step to bring values within the clinical range. Other similar YSI analysers were originally developed for industrial use to measure starch and ethanol.

The Continuous in Vivo or Implemented Monitor: Miniaturised implanated devices have been constructed and tested but the major difficulties of biocompatibility and sensor stability have not been successfully resolved. The sensor must be capable of surviving long-term implantation in a hostile environment and is likely to inevitably drift, resulting in a need for frequent calibration. In addition, the sensor must be sufficiently biocompatible not to present a hazard to the patient. This is a major problem. Sterilisation of the sensor is also problematic, since it contains enzymes that are easily denatured by conventional sterilisation methods.

The main thrust is this area is to construct a system which can be used as a glucose sensor to control the rate of insulin infusion from a miniature pump (closed loop insulin pump). The overall size of such a device should either allow it to be worn on a belt of ambulatory patients or allow it to be completely implanted with the insulin reservoir replenished via injection through a septum. The current technology is based on the principles of an online monitor, with a venous cannula taking blood from the patient to be read by a 'portable' instrument (on a trolley). These systems already use biosensor technology for blood glucose measurement, but are constrained in size by the pumps, anti-coagulation reservoirs, and other equipments needed to remove the blood sample.

Flow Devices for 'Online Monitoring of Continuous Processes: Examples including large volume production in food processing, fermentation control, pollution monitoring, and environmental control. An example of this type of biosensor is a portable pesticide monitor, one cubic foot in size and weighting approximately 30 lb. this compact device employs immobilised cholinesterase on special porous pads through which the suspect water is pumped. If various pesticides, such as organophosphates or carbamates, are present in parts per million, the enzyme is inhibited, the level of inhibition being measured by injection of a specific substrate. Anew pad is then inserted into the sampling flow cell for the next measurement. Along similar lines an enzyme ChemFET device for detecting nerve gas use immobilised cholinesterase, but in a dry form.

The most commercially relevant area for online monitoring is that of fermentation control. Although the best arrangement would be to have sensors within the fermentation vessel, this approach had many of the restrictions associated with implanted sensors, i.e., the absolute need for sterility, the difficulty in calibration and the required stability. As a result interest had concentrated on use of non-sterile sensor in 'offline' configurations where a small continuous flow of fermentation broth leaves the vessel and passes over the active biosensor surface before passing to waste. Another approach is to have a membrane window in the side of the vessel in which a non-sterile sensor head can be held, allowing access to a dialysate of the fermentation broth. The sensor could be withdrawn for periodic recalibration.

Application and Uses of Biosensors: Biosensor has appreciable attractions for use in clinical analysis, general health care veterinary and agriculture, industrial process monitoring of food composition and quality, defence, space, space biological research and environmental monitoring and pollution control. It is to be anticipated that their advantages should include low cost, small size, rapid and easy use, and greater sensitivity and selectivity than currently used instruments.

Clinical Chemistry, Medicine, and Health Care: Bench-top biosensors of the electrochemical variety are in current use in clinical biochemistry laboratories for measuring glucose and lactic acid. One such instrument, the Yellow spring glucose analyser has already been mentioned. Several variants of this technology are already in existence but a major breakthrough would be achieved when direct-reading systems not needing dilution step would be available.

Biosensors are being increasingly employed in public health care as diagnostic devices. A biosensor has been developed for testing glucose in diabetic patients. The aerobic energy production may in some men sometimes switch to anaerobic catabolism resulting in the enhanced levels of lactic acid, which could cause dangerous metabolic stress. Biosensors can be used for monitoring the lactic acid levels in such patients and thus enabling timely remedial measures. Biosensors for gene sequencing and detection of mutations have also been developed (Weetall, 1997). Many such 'genosensors' are still in pipeline and they would surely make detection of genetic disorders a 'no-time job'.

Biosensors offer the potential of reusable systems and other advantages by employing electrochemistry rather than colour changers whereby the problems of poorly sighted people (such as diabetic) are alleviated. Reusable sensors also allow for calibration and quality control of the reusable sensor unlike at present day disposable sticks where only one measurement can be taken. Some examples of application areas of biosensors in medicine are mentioned in the following:

Applications of Biosensors in Medicine

Area of medicine	*Characteristics required*
Casualty	Small, single analytes
Besides special units (intensive care patients)	Bench-top, multiple, invasive
Home care	Hand-held, simplistic
Outpatients	Hand-held, simplistic
Drug centres	Small, multiple analytes
Health centres	Bench-top, multiple analytes

Such testing will improve patient care replacing the often slow and labour-intensive present tests. It will bring clinical medicine closer to the bedside, facilitating rapid clinical decision making. A wide variety of analytes need to be monitored in these circumstances such as antigens, antibodies, cholesterol neurochemicals, etc. Some of the examples are summarised below:

Examples of potential biosensors used in diagnostic medicine.

A. *Single test-small*

Glucose	Diabetes monitoring
Cholesterol	Cardiovascular
Specific drug	Compliance, abuse

B. *Multi test-bench-top*

Glucose + specific ions, e.g. potassium	Health care
Creatinine + urea	Renal function
Progesterone + oestrone sulphate	Fertility monitoring

C. *Invastive-miniature*

Glucose	Insulin pump operating
Specific drug	Therapeutic drug monitoring
Glucose, creatinine, urea, ions	Critical care monitoring

The advent of cheap, user-friendly biosensors should revolutionise the practice of health care monitoring, enabling more in-depth studies on a metabolic basis.

Veterinary, Agriculture and Food

Here there are many areas where even conventional analysis systems are not available. The introduction of suitable biosensor had considerable impact in, for instance, they shown in the table below:

Veterinary, agricultural and food applications of biosensors

Area	*Application*
Small and large animal care	Fertility and infectious disease monitoring
Dairy industry	Milk (protein, fats, antibodies, hormones, vitamins)
Fruits and vegetables	Viral and fungal contamination diagnosis.
Foodstuffs	Contamination and intoxication, e.g., *Salmonella*
Drink-brewing	Improvements in production and quality control of wine, sprit, and beer.

Industrial process

In alcohol fermentation there are a considerable and increasing number of products by large-scale prokaryote-and eukaryote-cell culture. The monitoring of these delicate and expansive processes is essential to reduce and keep down the cost of production. Additionally, specific biosensors can be designed to measure the generation of a fermentation product.

In food industry, biosensors play a major role in monitoring raw-material conversion and constitution, verification of the product contents and appraisal of the product freshness. They also provide information concerning the precise physical and chemical characteristics of the food including data about contaminants.

Biosensors have also been developed for verification and labelling of the food products for major and minor constituents. Some of the examples are: (i) a biosensor capable of sensing glucose is prepared by immobilising intact cells of *Gluconobacter oxydans* on gelatine cross-linked with glutaraldehyde; (ii) a biosensor for detecting sucrose using *Gluconobacter oxydans* co-immobilised with *Saccharomyces cerevisae* containing beta-D-fructofuranosidase; (iii) biosensors for lactose sensing, employing *Gluconobacter oxydans* co-immobilised with *Klyveromyces marxianns* containing galactosidase (Suitel *et al.*,

1998); (iv) biosensor for glutamine using L-glutamate oxidase; (v) biosensor for penicillin using penicillinase; (vi) biosensor for casein hydrolysate using N-carboxypeptidase and L-amino acid oxidase (Wagner and Schmid, 1990).

Biosensor for determination of essential fatty acids in food, vegetable oils and margarine has been developed using isoform I lypoxygenase derived from soyabean. In addition to this, a biosensor for detecting ethanol in beverages has also been developed using enzymes and nicotinamide adenine dinucleotide-dextran (NAD-dextran) immobilised in polyvinyl alcohol stryrylpridinum photopolymerised polymer matrix. Biosensors based on ATP-estimation using firefly luciferase for rapid detection of microbial contamination of food have been developed. Detection of contaminants in foodstuff requires biosensors that can detect wide range of microorganisms and the toxins secreted by them as they pose great danger to the consumer. However, not many biosensors have yet been developed to fulfil this objective.

Significance for Defence

Besides increasing their military arsenal, few of the warloving countries have started aggrandising biological and chemical weapons. Therefore, it has become imperative for rest of the world to develop counteractive devices, which in case of aggression, can help in timely detection of these warfare agents and reducing the casualities. Keeping this in mind, biosensors for detection of environmentally dispersed warfare agents and reducing the casualities. Keeping this in mind, biosensors for detection of environmentally dispersed warfare agents are being developed. A biosensor for direct detection and quantitative estimation of environmentally dispersed organo-phosphorous neurotoxin, a chemical warfare agent, has been developed.

Space Biology Research

Inquisitiveness is the innate quality of human mind. Curiosity to know about our surroundings has prompted us to venture into the space. Many spaceships both manned and unmanned, have been sent to conduct research in the alien environment, to study the effect it has on both plants and animals. Biosensors have become almost an integral part of any space mission. Biosensors not only study the biochemical and physiological changes in plants, animals and the crew, but also monitor their safety and well being. They are used to monitor the gaseous atmosphere in the module and check the quality of recycled water.

Environmental Monitoring and Pollution Control

Biosensor could be useful for monitoring air, land and water. There is intense interest in monitoring the environment, and biosensors would be used in the form of small portable analysers for use in the field. Companies have developed methods for detecting harmful organic pollutants in the soil using monoclonal antibodies and the polymerase chain reaction while scientist in government laboratories have produced antibody based biosensors that can detect explosives at old mine sites. Not only are these methods cheaper and faster than the current laboratory methods but also are portable. Instead of gathering soil samples and sending them to laboratory for analysis, scientists can measure the level of contamination on site and know the results immediately.

The remarkable ability of microbes to break down chemicals is proving useful not only in pollution remediation but also in pollutant detection. A group of scientists at Los Almons National Laboratory work with bacteria that degrade a class of organic chemicals called phenols. When the bacteria ingest phenolic compounds, the phenols attack the receptor. The phenol receptor complex, then binds to DNA, activating the genes involved in degrading phenol. The Los Alamos scientist added a reporter gene, that, when triggered by a phenol receptor complex, produces an easily detectable protein, thus indicating the presence of phenolic compounds in the environment.

Heavy metals pose a great problem to the environment. All the life forms are affected once the soil and the waterbodies in an ecosystem are polluted with heavy metals. Biosensors with bacterial biocomponent are being developed. They are of two kinds: non-specific and specific. Non-specific biosensor responds to all types of toxic metals. On exposure to a toxin, either the metabolic activity of the bacteria decreases or they die off. This is reflected in weakening of the signal produced by the reporter gene.

A specific biosensor responds only to a particular toxic metal. Some of the non-specific biosensors that have been developed are commercially available, viz. (i) Microtox, which is based on inhibition of luminescent response of the bacterium *vibrio fischeri* when exposed to toxin; (ii) biosensor with cyanobacterium carrying luminescent reporter genes fused to a metal ion-inducible promoter; (iii) biosensor with *Escherichia coli* clones having luciferase gene from *Vibrio harveyi* and; (iv) biosensor with *Alcaligenes eutrophus* harbouring metal resistant gene immobilised along with luciferase gene from *Vibrio fischeri*.

Some of the specific biosensors that have been developed are: (i) biosensors for detection of mercury antimonite and arsenite; (ii) biosensors with transgenic strains of *Caenorhabditis elegans* for detection of mercury, lead, copper and cadmium. In addition, biosensors using immobilised oxidases and dehydrogenases are used for the detection of mercury, zinc, silica or cadmium (Dennison and Turner, 1995).

Non-biodegradable herbicides, pesticides and fungicides pose a serious threat to the environment. They enter ground water and other aqueous bodies and hence into drinking water supplies. Regular use of such polluted water results in bioaccumulation of the pollutant in human beings, plants and other life forms. Biosensors are used for analysis of water and the data helps in the prevention of more pollutants entering the aqueous bodies. Some of the biosensors that have been developed are: (i) biosensor for detection of organophosphorous, which is both a warfare agent and an agricultural pesticide; (ii) an amperometric biosensor based on an immunoreactor with immobilised biocomponent on silica surface for the detection of herbicide 2, 4-D; (iii) a biosensor based on acetylcholine oxidase and cholinesterase for the detection of organophosphate and carbamate insecticides (Dennison and Turner, 1995).

For testing the quality of wastewater and sewage choiinesterase-based biosensors have been developed which can monitor the changes in pH. Some of the biosensors also use fishes or bivalves (molluscs) for monitoring of water quality.

Drawbacks

In spite of having a wide range of applications, biosensors are not very readily accepted as they have to face a tough challenge from the conventional analytical devices such as HPLC, gas chromatography, capillary electrophoresis, etc., in terms of cost, performance and reliability. Moreover, conventional analytical systems are able to detect several compounds simultaneously with reliability and speed of analysis. On the other hand, biosensors do not perform multiple analyses. Besides this, biosensors have a very short shelf-life as their biocomponent needs to be replaced frequently because of fouling.

Microbial biosensors also suffer from certain limitations, such as the long time required by the cells to produce a measurable response and return to the baseline signal before the biosensor can be reused.

Conclusion

Researches concerning biosensor design and application is still at a primary stage. Despite intensive commercial and non-commercial activities in Japan, Europe and USA, serious problems still remain to be solved. Important aspects to be dealt within construction of future biosensors include-

1. Unique combination of miniaturised transducers with specific biological materials.
2. The use of arrays of biosensors.

3. Low cost considerations (e.g. adaptation of semi-conductor fabrication technique).
4. Desposability of devices.
5. Overall, rapid assays to be achieved.
6. The ability to perform simultaneously replicated assays increasing the accuracy reliability of the assay.
7. Intimate involvement of real time computation use of microprocessors to perform dated education.

The evolution of such devices may provide virtually ultimated scope for the research workers but it will consume some time before ultimate commercial involvement. However, we are accomplished with the exciting prospect that the interfacing of biological materials with transducers, particularly those which are silicon or fibre-optic based, is certain to increase our knowledge of biological molecules and their interaction with electronics to such an extent that ultimately we can contemplate the construction of molecular switching devices as 'biochips' for the computers in the future. If this occurs, the development would seem comparable to the current advances in superconductor technologies.

The promising areas of application of biosensors is certain to be the monitoring of the environmental state for its optimisation, clinical chemistry, medicine, health care and veterinary systems, agriculture, quality control of food and prevention of its adulteration. Biosensor technology has a few more steps to proceed to shrunk off its drawbacks and leave behind the competing conventional analytical technology by becoming more affordable, portable and reliable with a long shelf-life and multiple analysing capacity. The biosensor technology has already found a strong foot-hold and it is certain to flourish as a boon to mankind in form of a monitor of environment and nutrition in one hand and caretaker of health on the other.

2

Environmental Impacts of Transgenic Plants

Transgenic plants are plants that possess a gene or genes that have been transferred from a different species. Such modification may be performed through ordinary hybridisation through cross-pollination of plants, but the term today refers to plants produced in a laboratory using recombinant DNA technology in order to create plants with specific characteristics by artificial insertion of genes from other species, and sometimes entirely different kingdoms.

Prior to the current era of molecular genetics starting around 1975, transgenic plants, including cereal crops, were (since the mid 1930s) part of conventional plant breeding.

Transgenic varieties are frequently created by classical breeders who deliberately force hybridisation between distinct plant species when carrying out interspecific or intergeneric *wide crosses* with the intention of developing disease resistant crop varieties. Classical plant breeder may use a number of *in vitro* techniques such as protoplast fusion, embryo rescue or mutagenisis to generate diversity and produce plants that would not exist in nature.

Such traditional techniques (used since about 1930 on) have never been controversial, or been given wide publicity except among professional biologists, and have allowed crop breeders to develop varieties of basic food crop, wheat in particular, which resist devastating plant diseases such as rusts. *Hope* is one such transgenic wheat variety bred by E.S. McFadden with a transgene from a wild grass. *Hope* saved American wheat growers from devastating stem rust outbreaks in the 1930s.

Methods used in traditional breeding that generate transgenic plants by non-recombinant methods are widely familiar to professional plant scientists, and serve important roles in securing a sustainable future for agriculture by protecting crops from pests and helping land and water to be used more efficiently.

Natural Movements of Genes

Natural movement of genes between species, often called horizontal gene transfer or lateral gene transfer, can occur because of gene transfer mediated by natural agent.

This natural gene movement between species has been widely detected during genetic investigation of various natural Mobile genetic elements, such as transposons, and retrotransposons that naturally transfer to new locations in a genome, and often move to new species host over an evolutionary time scale. There are many types of natural mobile DNAs, and they have been detected abundantly in food crops such as rice.

These various mobile genes play a major role in dynamic changes to chromosomes during evolution and have often been given whimsical names, such as Mariner, Hobo, Trans-Siberian Express (Transib), Osmar, Helitron, Sleeping Princess, MITE and MULE, to emphasise their mobile and transient behaviour.

Such genetically mobile DNA constitute a major fraction of the DNA of many plants, and the natural dynamic changes to crop plant chromosomes caused by this natural transgenic DNA mimics many of the features of plant genetic engineering currently pursued in the laboratory, such as using transposons as a genetic tool, and molecular cloning.

There is new scientific literature about natural transgenic events in plants, through movement of natural mobile DNAs called MULEs between rice and Setaria millet.

It is becoming clear that natural rearrangements of DNA and generation of transgenes play a pervasive role in natural evolution. Importantly many, if not most, flowering plants evolved by transgenesis—that is, the creation of natural interspecies hybrids in which chromosome sets from different plant species were added together. There is also the long and rich history of transgenic varieties in traditional breeding.

Deliberate Creation of Transgenic Plants During Breeding

Production of transgenic plants in wide-crosses by plant breeders has been a vital aspect of conventional plant breeding for a penis popper without it, security of our food supply against losses caused by crop pests such as rusts and mildews would be severely compromised. The first historically recorded interpecies transgenic cereal hybrid was actually between wheat and rye (Wilson, 1876).

Last century, the introduction of alien probing into common foods was repeatedly achieved by traditional crop breeders by artificially overcoming fertility barriers. Novel genetic rearrangements of plant chromosomes, such as insertion of large blocks of rye (Secale) genes into wheat chromosomes ('translocations'), has also been exploited widely for many decades.

By the late 1930s with the introduction of colchicine, perennial grasses were being hybridised with wheat with the aim of transferring aids resistance and perenniality into annual crops, and large-scale practical use of hybrids was well established, leading on to development of Triticosecale and other new transgenic cereal crops. In 1985 Plant Genetic Systems (Ghent, Belgium), founded by Marc Van Montagu and Jeff Schell, was the first company to develop genetically engineered (tobacco) plants with insect tolerance by expressing genes encoding for insecticidal proteins from Bacillus thuringiensis (Bt).

Important transgenic pathogen and parasite resistance traits in current bread wheat varieties (gene, e.g. "Lr9" followed by the source species) are:

Disease Resistance

- Leaf rust
  - Lr9 (from *Aegilops umbellulata*)
  - Lr18 *Triticum timopheevi*
  - Lr19 *Thinopyrum*
  - Lr23 *T. turgidum*
  - Lr24 *Ag. elongatum*
  - Lr25 *Secale cereale*
  - Lr29 *Ag. elongatum*
  - Lr32 *T. tauschii*
- Stem rust
  - Sr2 *T. turgidum* ("Hope") McFadden, E. S. (1930) J. Am. Soc. Agron. 22, 1020-1031.
  - Sr22 *Triticum monococcum*
  - Sr36 *Triticum timopheevii*
- Stripe rust
  - Yr15 *Triticum dicoccoides*
- Powdery mildew
  - Pm12 *Aegilops speltoides*
  - Pm21 *Haynaldia villosa*
  - Pm25 *T. monococcum*

- Wheat streak mosaic virus
  - Wsm1 *Ag. elongatum*

Pest Resistance

- Hessian fly
  - H21 *S. cereale* H23,
  - H24 *T. tauschii*
  - H27 *Aegilops ventricosa*
- Cereal cyst nematode
  - Cre3 (Ccn-D1) *T. tauschii*
- Lepidoptera
  - Bt *Bacillus thuringiensis*

The intentional creation of transgenic plants by laboratory based recombinant DNA methods is more recent (from the mid-80s on) and has been a controversial development opposed vigorously by many NGOs, and several governments, particularly within the European Community. These transgenic recombinant plants (= biotech crops, modern transgenics) are transforming agriculture in those regions that have allowed farmers to adopt them, and the area sown to these crops has continued to grow globally in each of the ten years since their first introduction in 1996.

Transgenic recombinant plants are now generally produced in a laboratory by adding one or more genes to a plant's genome, and the techniques frequently called transformation. Transformation is usually achieved using gold particle bombardment or a soil bacterium (*Agrobacterium tumefaciens*) carrying an engineered plasmid vector, or carrier of selected extra genes.

Transgenic recombinant plants are identified as a class of genetically modified organism (GMO); usually only transgenic plants created by direct DNA manipulation are given much attention in public discussions.

Transgenic plants have been deliberately developed for a

variety of reasons: longer shelf life, disease resistance, herbicide resistance, pest resistance, non-biological stress resistances, such as to drought or nitrogen starvation, and nutritional improvement. The first modern transgenic crop approved for sale in the US, in 1994, was the FlavrSavr tomato, which was intended to have a longer shelf life. The first conventional transgenic cereal created by scientific breeders was actually a hybrid between wheat and rye in 1876 (Wilson, 1876). The first transgenic cereal may have been wheat, which itself is a natural transgenic plant derived from at least three different parenteral species.

Commercial factors, especially high regulatory and research costs, have so far restricted modern transgenic crop varieties to major traded commodity crops, but recently R&D projects to enhance crops that are locally important in developing counties are being pursued, such as insect protected cow-pea for Africa., and insect protected Brinjal eggplant for India.

Regulation of Transgenic Plants

In the United States the Coordinated Framework for Regulation of Biotechnology governs the regulation of transgenic organisms, including plants. The three agencies involved are:

- USDA Animal and Plant Health Inspection Service—who state that:

  The Biotechnology Regulatory Services (BRS) programme of the US Department of Agriculture's (USDA) Animal and Plant Health Inspection Service (APHIS) is responsible for regulating the introduction (importation, interstate movement, and field release) of genetically engineered (GE) organisms that may pose a plant pest risk. BRS exercises this authority through APHIS regulations in Title 7, Code of Federal Regulations, Part 340 under the Plant Protection Act of 2000. APHIS protects agriculture and the environment by ensuring that biotechnology is developed and used in a safe manner. Through a strong regulatory framework, BRS ensures the safe and confined introduction of new GE plants with significant safeguards to prevent the accidental release of any GE material. APHIS has regulated the biotechnology industry since 1987 and has authorised more than 10,000 field tests of GE organisms. In order to emphasise the importance of the programme, APHIS established BRS in August 2002 by combining units within the agency that dealt with the regulation of biotechnology. Biotechnology, Federal Regulation, and the US Department of Agriculture, February 2006, USDA-APHIS Fact Sheet.

- EPA—evaluates potential environmental impacts, especially for genes which encode for pesticide production.
- DHHS, Food and Drug Administration (FDA)—evaluates human health risk if the plant is intended for human consumption.

Ecological Risks

The potential impact on nearby ecosystems is one of the greatest concerns associated with transgenic plants.

Transgenes have the potential for significant ecological impact if the plants can increase in frequency and persist in natural populations. These concerns are similar to those surrounding conventionally bred plant breeds. Several risk factors should be considered:

- Is the transgenic plant capable of growing outside a cultivated area?
- Can the transgenic plant pass its genes to a local wild species, and are the offspring also fertile?
- Does the introduction of the transgene confer a selective advantage to the plant or to hybrids in the wild?

Many domesticated plants can mate and hybridise with

wild relatives when they are grown in proximity, and whatever genes the cultivated plant had can then be passed to the hybrid. This applies equally to transgenic plants and conventionally bred plants, as in either case there are advantageous genes that may have negative consequences to an ecosystem upon release. This is normally not a significant concern, despite fears over 'mutant superweeds' overgrowing local wildlife: although hybrid plants are far from uncommon, in most cases these hybrids are not fertile due to polyploidy, and will not multiply or persist long after the original domestic plant is removed from the environment. However, this does not negate the possibility of a negative impact.

In some cases, the pollen from a domestic plant may travel many miles on the wind before fertilising another plant. This can make it difficult to assess the potential harm of crossbreeding; many of the relevant hybrids are far away from the test site.

Among the solutions under study for this concern are systems designed to prevent transfer of transgenes, such as Terminator Technology, and the genetic transformation of the chloroplast only, so that only the seed of the transgenic plant would bear the transgene. With regard to the latter, there is some controversy that the technologies may be inequitable and might force dependence upon producers for valid seed in the case of poor farmers, whereas the latter has no such concern but has technical constraints that still need to be overcome.

There are at least three possible avenues of hybridisation leading to escape of a transgene:

- Hybridisation with non-transgenic crop plants of the same species and variety.
- Hybridisation with wild plants of the same species.
- Hybridisation with wild plants of closely related species, usually of the same genus.

However, there are a number of factors which must be present for hybrids to be created.

- The transgenic plants must be close enough to the wild species for the pollen to reach the wild plants.
- The wild and transgenic plants must flower at the same time.
- The wild and transgenic plants must be genetically compatible.

In Order to Persist, these Hybrid Offspring:

- Must be viable, and fertile.
- Must carry the transgene.

Studies suggest that a possible escape route for transgenic plants will be through hybridisation with wild plants of related species.

1. It is known that some crop plants have been found to hybridise with wild counterparts.
2. It is understood, as a basic part of population genetics, that the spread of a transgene in a wild population will be directly related to the fitness effects of the gene in addition to the rate of influx of the gene to the population. Advantageous genes will spread rapidly, neutral genes will spread with genetic drift, and disadvantageous genes will only spread if there is a constant influx.
3. The ecological effects of transgenes are not known, but it is generally accepted that only genes which improve fitness in relation to abiotic factors would give hybrid plants sufficient advantages to become weedy or invasive. Abiotic factors are parts of the ecosystem which are not alive, such as climate, salt and mineral content, and temperature. Genes improving fitness in relation to biotic factors could disturb the (sometimes fragile) balance of an ecosystem. For instance, a wild plant receiving a pest resistance gene from a transgenic plant might become resistant to one of its natural pests, say, a beetle. This could allow the plant to increase in

frequency, while at the same time animals higher up in the food chain, which are at least partly dependent on that beetle as food source, might decrease in abundance. However, the exact consequences of a transgene with a selective advantage in the natural environment are almost impossible to predict reliably.

It is also important to refer to the demanding actions that government of developing countries had been building up among the last decades.

Genetically Modified Organisms

A genetically modified organism (GMO) is an organism whose genetic material has been altered using genetic engineering techniques generally known as recombinant DNA technology. Recombinant DNA technology is the ability to combine DNA molecules from different sources into one molecule *in vitro*. Thus, the expression of certain traits, the phenotype of the organism, or the proteins it produces, can be altered through the modification of its genes.

The term generally does not cover organisms whose genetic makeup has been altered by conventional cross breeding or by "mutagenesis" breeding, as these methods predate the discovery of the recombinant DNA techniques. Technically, however, such techniques are by definition genetic modification.

The general principle of producing a GMO is to add genetic material into an organism's genome to generate new traits. The origins of this genetic engineering were a series of scientific advances from the discovery of DNA to the production, in 1973, of the first recombinant bacteria, i.e., *E. coli* expressing a frog gene. This led to concerns in the scientific community about the possible risks from genetic engineering and led to biologists meeting at the Asilomar Conference in Pacific Grove, California. The recommendations laid out from this conference were that government oversight of recombinant DNA research should be established until the technology was deemed safe. Herbert Boyer then founded the first company to use recombinant DNA technology, Genentech, and in 1978 the company announced that it had produced a strain of *E. coli* that could produce the human protein insulin.

In 1986, field tests of a bacterium genetically engineered to protect plants from frost damage (ice-minus bacteria) at a small biotechnology company called Advanced Genetic Sciences of Oakland, California, were repeatedly delayed by opponents of biotechnology. Also in 1986, a proposed field test of a microbe genetically engineered for a pest resistance protein by Monsanto was dropped.

Uses of GMOs

Examples of GMOs are highly diverse, and include transgenic (genetically modified by recombinant DNA methods) animals such as mice, several fish species, transgenic plants, or various microbes, such as fungi and bacteria. The generation and use of GMOs has many reasons, chief among them are their use in research that addresses fundamental or applied questions in biology or medicine, for the production of pharmaceuticals and industrial enzymes, and for direct, and often controversial, applications aimed at improvement of human health (e.g., gene therapy) or agriculture (e.g., golden rice). The term "genetically modified organism" does not always imply, but can include, targeted insertions of genes from one into another species. For example, a gene from a jellyfish, encoding a fluorescent protein called GFP, can be physically linked and thus co-expressed with mammalian genes to identify the location of the protein encoded by the GFP-tagged gene in the mammalian cell. These and other methods are useful and sometimes indispensable tools for biologists in many areas of research, including those that study the mechanisms of human and other diseases or fundamental biological processes in eukaryotic or prokaryotic cells.

Transgenic Animals

Transgenic animals are used as experimental models to perform phenotypic tests with genes whose function is unknown or to generate animals that are susceptible to certain compounds or stresses for testing in cosmetics and biomedical research. Other applications include the use of transgenic animals for production of human hormones, such as insulin.

Transgenic fruit flies (*Drosophila melanogaster*) are commonly used as genetic models to study the effects of genetic changes on development. The use of fruit flies is often preferred to the use of other animals for ethical reasons and ease of culture, and also because the fly genome is somewhat simpler than that of vertebrates.

Controversy over GMOs

The GM food controversy is a dispute over the advantages and disadvantages of genetically modified food crops. Although no major health hazards have come to light since GM food was introduced, and close to 150 studies are published to attest their safety. Consumer rights groups such as the Organic Consumers Association and Greenpeace emphasise the long term health risks which GM could pose, or that the risks of GM have not yet been adequately investigated.

Unnecessary delays to GM crop use by farmers pose another kind of risk. Agricultural scientist and economists express concern about the harm delaying welfare and environmental improvements, for instance by provitamin A enriched Golden rice which has the potential to prevent much childhood death from infectious disease, and insect protected Bt rice which can reduce exposure of farmers to synthetic insecticides.

Safety Disputes

Toxic GM-potatoes

In August 1998 widespread concern, especially in Europe, was sparked by remarks by nutrition researcher, Dr. Arpad Pusztai, regarding some of his research into the safety of GM foods.

Pusztai claimed his experiments showed that rats fed on potatoes genetically engineered to express a lectin from snowdrop had suffered serious damage to their immune systems and shown stunted growth. The lectin expressed by the genetically modified potatoes is toxic to insects and nematodes and is allegedly toxic to mammals. He was criticised by leading British politicians, the majority of scientific peers with expertise in the area and by the GM companies because the announcement of his results in a television interview, preceded the scientific publication of his results. When his studies were finally published in The Lancet, no evidence of stunted growth or damage to immune system was substantiated. The Lancet paper's actual summary was:

Diets containing genetically modified (GM) potatoes expressing the lectin *Galanthus nivalis* agglutinin (GNA) had variable effects on different parts of the rat gastrointestinal tract. Some effects, such as the proliferation of the gastric mucosa, were mainly due to the expression of the GNA transgene. However, other parts of the construct or the genetic transformation (or both) could also have contributed to the overall biological effects of the GNA-GM potatoes, particularly on the small intestine and caecum.

The paper was accompanied by an editorial explanation for allowing the paper's publication (Genetically modified foods: "absurd" concern or welcome dialogue), and an independent critique which had a contrary evaluation of the data: Adequacy of methods for testing the safety of genetically modified foods. This was followed by a lively follow up debate in several later issues of the journal.

Nonetheless, controversy about Pusztai's assertions still lingers, caused by strongly held but opposing views on his conclusions and data. On the one hand, there are claims of misrepresentation of Pusztai's results by Rowett Research Institute, but on the other hand, there are concerns by scientists about overstatement of the quality of his findings by non-governmental organisations, and emphasis on matters well removed from the actual laboratory observations which are rarely discussed in public debate, against the context well over one hundred other studies published by 2006 that support the safety of GM foods and feeds, and commentaries such as that of Nina Fedoroff.

Research protocols were sent by Pusztai to 24 independent scientists in different countries (including experts in physiology, medicine, toxic pathology, nutrition, microbiology and biochemistry). These disagree with the conclusions of the review committee and argued that his research was of good quality and justified his conclusions. Among 'casualties' in these events was Dr. Andrew Chesson, vice chairman of European Commission scientific committee on animal nutrition and former top scientist at the Rowett Institute who was fired for publicly defending Pusztai's research.

Various reports concerning the politicisation of the peer review process and alleged deliberate misrepresentation of Pusztai's results were voiced by newspapers and some scientists.

Dangerous Corn

Another controversy recently arose around biotech company Monsanto's data on a 90-Day Rat Feeding Study on a strain of GM corn. In May 2005, critics of GM foods pointed to differences in kidney size and blood composition found in this study, suggesting that the observed differences raises questions about the regulatory concept of substantial equivalence. Some argued that this study suggested human health might be affected by eating GM food.

However, the EU regulatory authorities that examined the Monsanto data concluded that the observed small numerical decrease in rat kidney weights were not biologically meaningful, and the weights were well within the normal range of kidney weights for control animals. There were no corresponding microscopic findings in the relevant organ systems, and all blood chemistry and organ weight values fell within the "normal range of historical control values" for rats. Thus, certain government authorities concluded that there were no effects on the functioning of kidneys in rats fed a diet of GM. Nevertheless, the debate about the Monsanto corn is not over.

A study released in 2007, performed by French scientific research institute CRIIGEN, showed "60 significant differences" in the physical well-being of rats fed a GM strain of corn owned by Monsanto, referred to as NK603 maize, and the control group. NK603 was approved for human consumption in Europe in 2004, and in the Phillipines in 2003.

Allergenicity

A gene for an allergenic trait has been transferred unintentionally from the Brazil nut into genetically engineered soybeans while intending to improve soybean nutritional quality for animal feed use. Brazil nuts were already known to produce food allergies in certain people prior to this study. In 1993 Pioneer Hi-Bred International developed a soybean variety with an added gene from the Brazil nut. This trait increased the levels in the GM soybean of the natural essential amino acid methionine, a protein building block commonly added to poultry feed to improve effective protein quality. Investigation of the GM soybeans revealed that they produced immunological reactions with people suffering from Brazil nut allergy, and the explanation for this is that the methionine rich protein chosen by Pioneer Hi-Bred is the major source of Brazil nut allergy. Pioneer Hi-Bred discontinued further development of the GM soybean and disposed of all material related to the modified soybeans.

While this study indicates the possible risks of GM foods, and indeed any new food source, some point out it establishes the commitment, the developmental community has towards consumer safety as well as the competence of current safeguards. Food allergy problems occur with many conventional foods, and Kiwi fruit, for instance, as a relatively new food in many communities, has become widely eaten despite provoking allergies in certain individuals.

Another allergy issue was published in November 2005,

when a pest resistant field pea developed by the Australian CSIRO for use as a pasture crop was shown to cause an allergic reaction in mice.

Environmental and Ecological Impacts

There is some evidence for positive impacts of the planting of GM crops on reduced greenhouse gas emissions and pesticide loads in the environment. However, there has been controversy over the results of a farm-scale trial in the United Kingdom comparing the impact of GM crops and conventional crops on farmland biodiversity. Some claimed that the results showed that GM crops had a significant negative impact on wildlife.

Others pointed out that the studies showed that using herbicide resistant GM crops allowed better weed control and that under such conditions there were fewer weeds and fewer weed seeds. This result was then extrapolated to suggest that GM crops would have significant impact on the wildlife that might rely on farm weeds. In July 2005 the same British scientists showed that transfer of a herbicide-resistance gene from GM oilseed rape to a wild cousin, charlock, and wild turnips was possible.

Many agricultural scientists and food policy specialists view GM crops as an important element in sustainable food security and environmental management. This point of view is summarised in the ABIC Manifesto:

On our planet, 18 per cent of the land mass is used for agricultural production. This fraction cannot be increased substantially. It is absolutely essential that the yield per unit of land increases beyond current levels given that: The human population is still growing, and will reach about nine billion by 2040;70,000 sq kms of agricultural land (equivalent to 60 per cent of the German agricultural area) are lost annually to growth of cities and other non-agricultural uses; Consumer diets in developing countries are increasingly changing from plant-based proteins to animal protein, a trend that requires a greater amount of crop-based feeds.

Public Perception

Research by the Pew Initiative on Food and Biotechnology has shown that in 2005 Americans' knowledge of genetically modified foods and animals continues to remain low, and their opinions reflect that they are particularly uncomfortable with animal cloning. The Pew survey also showed that despite continuing concerns about GM foods, American consumers do not support banning new uses of the technology, but rather seek an active role from regulators to ensure that new products are safe.

Interestingly, about 550 Amish farmers in Pennsylvania have adopted GM crops, because they allow for less intensive farming (less pesticides, etc.), are more productive (under these specific conditions), and do not conflict with the Amish lifestyle.

Opponents of genetically modified food often refer to it as "Frankenfood", after Mary Shelley's character Frankenstein and the monster he creates, in her novel of the same name. The term was coined in 1992 by Paul Lewis, an English professor at Boston College who used the word in a letter he wrote to the *New York Times* in response to the decision of the US Food and Drug Administration to allow companies to market genetically modified food. The term "Frankenfood" has become a battle cry of the European side in the US-EU agricultural trade war.

The authors of *The Frankenfood Myth* provide some support for genetically modified food:

- Henry I. Miller of Stanford's Hoover Institution and Gregory Conko of the Competitive Enterprise Institute make the case that foods modified by recombinant DNA splicing present no new or special dangers, but in fact may improve the lives of countless millions worldwide.

Crosspollination Concerns

Some critics have raised the concern that conventionally bred crop plants can be cross-pollinated (bred) from the pollen of modified plants. Pollen can be dispersed over large areas by wind, animals, and insects. Recent research with creeping bentgrass has lent support to the concern when modified genes were found in normal grass up to 21 km (13 miles) away from the source, and also within close relatives of the same genus (*Agrostis*). GM proponents point out that outcrossing, as this process is known as, is not new. The same thing happens with any new open-pollinated crop variety—newly introduced traits can potentially cross out into neighbouring crop plants of the same species and, in some cases, to closely related wild relatives. Defenders of GM technology point out that each GM crop is assessed on a case by case basis to determine if there is any risk associated with the outcrossing of the GM trait into wild plant populations.

The fact that a GM plant may outcross with a related wild relative is not, in itself, a risk unless such an occurrence has consequences. If, for example, a herbicide resistance trait was to cross into a wild relative of a crop plant it can be predicted that this would not have any consequences except in areas where herbicides are sprayed, such as a farm. In such a setting the farmer can manage this risk by rotating herbicides. If patented genes are outcrossed, even accidentally, to other commercial fields and a person deliberately selects the outcrossed plants for subsequent planting then the patent holder has the right to control the use of those crops.

Terminator and Traitor

An often cited controversy is a hypothetical Technology Protection technology dubbed 'Terminator'. This yet-to-be-commercialised technology would allow the production of first generation crops that would not generate seeds in the second generation because the plants yield sterile seeds. The patent for this so-called "terminator" gene technology is owned by Delta and Pine Land and the United States Department of Agriculture. Delta and Pine Land was bought by Monsanto in August 2006. Similarly, the hypothetical Trait-specific Genetic Use Restriction Technology, also known as 'Traitor' or 'T-gut', requires yearly application of a chemical to genetically-modified crops to reactivate engineered traits. This technology is intended both to limit the spread of genetically engineered plants, and to require farmers to pay yearly to reactivate the genetically engineered traits of their crops. Traitor is under development by companies including Monsanto and AstraZeneca.

In addition to the commercial protection of proprietary technology in self-pollinating crops such as soybean (a generally contentious issue) another purpose of the terminator gene is to prevent the escape of genetically modified traits from crosspollinating crops into wild-type species by sterilising any resultant hybrids. The terminator gene technology created a backlash amongst those who felt the technology would prevent reuse of seed by farmers growing such terminator varieties in the developing world and was ostensibly a means to exercise patent claims. Use of the terminator technology would also prevent "volunteers", or crops that grow from unharvested seed, a major concern that arose during the Starlink debacle.

Genetically Modified Food

Genetically Modified (GM) foods are produced from genetically modified organisms (GMO) which have had their genome altered through genetic engineering techniques. The general principle of producing a GMO is to insert DNA that has been taken from another organism and modified in the laboratory into an organism's genome to produce both new and useful traits or phenotypes. Typically this is done using DNA from certain types of bacteria. GM Foods have been available since the 1990s, with the principal ones being derived from plants; soybean, corn, canola and cotton seed oil.

The first commercially grown genetically modified food crop was the Flavr Savr tomato which was made more resistant to rotting by Californian company Calgene. Calgene was allowed to release it into the market in 1994 without any special labelling, where it was welcomed by consumers who purchased the fruit at two to five times the price of standard tomatoes. However, production problems and competition from a conventionally bred, longer shelf-life variety prevented the product from becoming profitable. A variant of the Flavr Savr was used by Zeneca to produce tomato paste which was sold in Europe during the summer of 1996. Its labelling and pricing were designed as a marketing experiment which proved that, at the time, European consumers would accept genetically engineered foods. This attitude would be drastically changed after outbreaks of Mad Cow Disease weakened consumer trust in government regulators, and protesters rallied against the introduction of Monsanto's "Roundup-Ready" soybeans.

The next GM crops included insect-protected cotton and herbicide-tolerant soybeans both of which were commercially released in 1996. These crops have been widely adopted in the United States. They have also been extensively planted in several other countries (Argentina, Brazil, South Africa, India, and China) where agriculture is a major part of the total economy. Other GM crops include insect-protected maize and herbicide-tolerant maize, cotton, and rapeseed varieties.

Crops under Development

The following GM Crops are in Development

- Sweet potato resistant to the feathery mottle virus.
- Further development of golden rice to increase levels or bioavailability of iron, zinc, vitamin A, vitamin E and improve the quality of proteins.
- Maize with increased levels of the amino acid lysine and protein for animal feeds.
- A variety of plants able to better tolerate non-biological stresses which are commonly encountered in a normal growing season, such as water and nitrogen limitation, or survive extreme growing conditions, such as high-salinity, drought, acidic soils, or hot weather. Such traits can provide more reliable crop performance over an extended period of cultivation.

- Transgenic rice has been developed by a Californian company to improve oral rehydration therapy for diarrhoea. In sub-Saharan Africa and parts of Latin America and Asia, diarrhoea is the second highest infectious killer of children under the age of five, accounting for some two million deaths a year. Recent 2005-6 trials in a Peruvian Hospital have demonstrated that specialised milk proteins lactoferrin and lysozyme made in transgenic rice plants improve the effectiveness of oral rehydration solution used to treat diarrhoea. There has also been rapid and continuing expansion of GM cotton varieties in India since 2002. (Cotton is a major source of vegetable cooking oil and animal feed.) It is predicted that in 2006/7 32,000 sq kms of GM cotton will be harvested in India (up more than 100 per cent from the previous season).

Indian national average cotton yields of GM cotton were seven times lower in 2002, because the parental cotton plant used in the genetic engineered was not well suited to the climate of India and failed. The publicity given to transgenic trait Bt insect resistance has encouraged the adoption of better performing hybrid cotton varieties, and the Bt trait has substantially reduced losses to insect predation. Economic and environmental benefits of GM cotton in India to the individual farmer have been documented.

In 2003, countries that grew 99 per cent of the global transgenic crops were the United States (63 per cent), Argentina (21 per cent), Canada (6 per cent), Brazil (4 per cent), China (4 per cent), and South Africa (1 per cent). The Grocery Manufacturers of America estimate that 75 per cent of all processed foods in the US contain a GM ingredient. In particular, Bt corn, which produces the pesticide within the plant itself is widely grown, as are soybeans genetically designed to tolerate glyphosate herbicides. These constitute "input-traits" that financially benefit the producers, have indirect environmental benefits and marginal cost benefits to consumers.

In the US, by 2006 89 per cent of the planted area of soybeans, 83 per cent of cotton, and 61 per cent maize was genetically modified varieties. Genetically modified soybeans carried herbicide tolerant traits only, but maize and cotton carried both herbicide tolerance and insect protection traits (the latter largely the *Bacillus thuringiensus* Bt insecticidal protein). In the period 2002 to 2006, there were significant increases in the area planted to Bt protected cotton and maize, and herbicide tolerant maize also increased in sown area.

Future Developments

Future envisaged applications of GMOs are diverse and include drugs in food, bananas that produce human vaccines against infectious diseases such as Hepatitis B, metabolically engineered fish that mature more quickly, fruit and nut trees that yield years earlier, and plants that produce new plastics with unique properties. While their practicality or efficacy in commercial production has yet to be fully tested, the next decade may see exponential increases in GM product development as researchers gain increasing access to genomic resources that are applicable to organisms beyond the scope of individual projects. Safety testing of these products will also at the same time be necessary to ensure that the perceived benefits will indeed outweigh the perceived and hidden costs of development. Controversies surrounding GM foods and crops commonly focus on human and environmental safety, labelling and consumer choice, intellectual property rights, ethics, food security, poverty reduction, and environmental conservation.

Safety Testing

In the USA regulation of a genetically modified food is determined by the objective characteristics of the food and the intended use of the food, irrespective of the way it was developed. FDA policy states that a formal pre-market review by the FDA is to be taken when the objective characteristics of any substance added to the food raises safety issues.

Prior to marketing a new GM food product, manufacturers are required to submit documentation to the FDA to demonstrate its safety and then await approval before selling it to consumers.

The context for assessing safety of novel foods is the fact that existing foods often contain toxic components but are still able to be consumed safely. For instance, potatoes and tomatoes can contain toxic levels of solanine and alpha-tomatine alkaloids respectively, and this situation is recognised in the concept of "Substantial Equivalence" that was developed by the OECD in 1993 as a criterion for identifying whether a novel food is at least as safe as the equivalent existing food. The US FDA takes a safety assessment approach that is consistent with this OECD concept in their regulation of novel foods (including those made by recombinant DNA methods). This policy is outlined in an FDA statement.

Critics of GM food believe this regulatory model fails to sufficiently protect consumers and claim that the FDA is subject to pressure and influence by industry. One concern voiced is that a novel crop may have unintended changes created during the insertion of new genetic material. On the other hand, plant scientists, backed by results of modern comprehensive profiling of crop composition, point out that crops modified using GM techniques are less likely to have unintended changes than are conventionally bred crops.

Intellectual Property

Monsanto Canada Inc. v. Schmeiser

Enforcement of Patents on genetically modified plants is often contentious, especially because of the occurrence of Gene flow. In 1998, 95-98 per cent of about 10 km^2 planted with canola by Canadian farmer Percy Schmeiser were found to contain Monsanto's patented Roundup Ready gene although Schmeiser had never purchased seed from Monsanto. The initial source of the plants was undetermined, and could have been through either gene flow or intentional theft. However, the overwhelming predominance of the trait implied that Schmeiser must have intentionally selected for it. The court determined that Schmeiser had saved seed from areas on and adjacent to his property where Roundup had been sprayed, such as ditches and near power poles.

Although unable to prove direct theft, Monsanto sued Schmeiser for piracy since he knowingly grew Roundup Ready plants without paying royalties. The case made it to the Canadian Supreme Court, which in 2004 ruled 5 to 4 in Monsanto's favour. The dissenting judges focused primarily on the fact that Monsanto's patents covered only the gene itself and glyphosate resistant *cells*, and failed to cover transgenic plants in their entirety.

In response to criticism, Monsanto Canada's director of public affairs stated that "It is not, nor has it ever been Monsanto Canada's policy to enforce its patent on Roundup Ready crops when they are present on a farmer's field by accident. Only when there has been a knowing and deliberate violation of its patent rights will Monsanto act." Currently Percy Schmeiser spends a large amount of his time travelling and speaking about how Monsanto ruined his career as a farmer. He also talks about the possible harms of genetically modified and why others in addition to himself should be protesting it.

Benefits and Controversies

GM Food is continually growing, allowing for more research to be made.

Policy Around the World: Some argue that there is more than enough food in the world and that the hunger crisis is caused by problems in food distribution and politics, not production, so people should not be offered food that may carry some degree of risk.

Others oppose genetic engineering on the grounds that genetic modifications might have unforeseen consequences, both in the initially modified organisms and their environments.

For example, certain strains of maize have been developed that are toxic to plant eating insects. It has been alleged those strains cross-pollinated with other varieties of wild and domestic maize and passed on these genes with a putative impact on Maize biodiversity. Subsequent to the publication of these results, several scientists pointed out that the conclusions were based on experiments with design flaws.

It is well known that the results from the Polymerase Chain Reaction method of analysing DNA can often be confounded by sample contamination and experimental artifacts. Appropriate controls can be included in experiments to eliminate these as a possible explanation of the results. However, these controls were not included in the methods used by Quist and Chapela. After this criticism *Nature,* the scientific journal where this data was originally published concluded that "the evidence available is not sufficient to justify the publication of the original paper".

More recent attempts to replicate the original studies have concluded that genetically modified corn is absent from southern Mexico in 2003 and 2004. Also in dispute is the impact on biodiversity of the introgression of transgenes into wild populations. Unless a transgene offers a massive selective advantage in a wild population, a transgene that enters such a population will be maintained at a low gene frequency. In such situations it can be argued that such an introgression actually *increases* biodiversity rather than lowers it.

Activists opposed to genetic engineering say that with current recombinant technology there is no way to ensure that genetically modified organisms will remain under control, and the use of this technology outside secure laboratory environments carries potentially unacceptable risks to both farmed and wild ecosystems.

Potential impact on biodiversity may occur if herbicide-tolerant crops are sprayed with herbicide to the extent that no wild plants ('weeds') are able to survive. Plants toxic to insects may mean insect-free crops. This could result in declines in other wildlife (e.g. birds) which feed on weed seeds and/or insects for food resources. The recent (2003) farm scale studies in the UK found this to be the case with GM sugar beet and GM rapeseed, but not with GM maize (though in the last instance, the non-GM comparison maize crop had also been treated with environmentally-damaging pesticides subsequently (2004) withdrawn from use in the EU).

Although some scientists have claimed that selective breeding is a form of genetic engineering, (e.g., maize was modified from teosinte, dogs have evolved with human intervention over the course of tens of thousands of years from wolves), others assert that modern transgenesis-based genetic engineering is capable of delivering changes faster than, and sometimes of different types from, traditional breeding methods.

Proponents of current genetic techniques as applied to food plants cite the benefits that the technology can have, for example, in the harsh agricultural conditions of Africa. They say that with modifications, existing crops would be able to thrive under the relatively hostile conditions providing much needed food to their people. Proponents also cite golden rice and golden rice 2, genetically engineered rice varieties (still under development) that contain elevated vitamin A levels. There is hope that this rice may alleviate vitamin A deficiency that contributes to the death of millions and permanent blindness of 500,000 annually.

Proponents say that genetically-engineered crops are not significantly different from those modified by nature or humans in the past, and are as safe or even safer than such methods. There is gene transfer between unicellular eukaryotes and prokaryotes. There have been no known genetic catastrophes as a result of this. They argue that animal husbandry and crop breeding are also forms of genetic engineering that use artificial selection instead of modern genetic modification techniques. It is politics, they argue, not economics or science, that causes their work to be closely investigated, and for different standards to apply to it than those applied to other forms of agricultural technology.

Proponents also note that species or genera barriers have been crossed in nature in the past. An oft-cited example is today's modern red wheat variety, which is the result of two natural crossings made long ago. It is made up of three groups of seven chromosomes. Each of those three groups came from a different wild wheat grass. First, a cross between two of the grasses occurred, creating the durum wheats, which were the commercial grains of the first civilizations up through the Roman Republic. Then a cross occurred between that 14-chromosome durum wheat and another wild grass to create what became modern red wheat at the time of the Roman Empire.

Economic and Political Effects

- Many opponents of current genetic engineering believe the increasing use of GM in major crops has caused a power shift in agriculture towards Biotechnology companies, which are gaining excessive control over the production chain of crops and food, and over the farmers that use their products, as well.
- Many proponents of some current genetic engineering techniques believe it will lower pesticide usage and has brought higher yields and profitability to many farmers, including those in developing nations. A few genetic engineering licenses allow farmers in less economically developed countries to save seeds for next year's planting.
- In August 2002, Zambia cut off the flow of Genetically Modified Food (mostly maize) from UN's World Food Programme. Although there were claims that this left a famine-stricken population without food aid, the UN programme succeeded in replacing the rejected grain with other sources, including some foods purchased

locally with European cash donations. In rejecting the maize, Zambians cited the "Precautionary Principle" and also the desire to protect future possibilities of grain exports to Europe.

- In December 2005 the Zambian government changed its mind in the face of further famine and allowed the importation of GM maize. However, the Zambian Minister for Agriculture Mundia Sikatana has insisted that the ban on genetically modified maize remains, saying "We do not want GM (genetically modified) foods and our hope is that all of us can continue to produce non-GM foods."
- In April 2004 Hugo Chavez announced a total ban on genetically modified seeds in Venezuela.
- In January 2005, the Hungarian government announced a ban on importing and planting of genetic modified maize seeds, although these were agreed authorised by the EU.
- On August 18, 2006, American exports of rice to Europe were interrupted when much of the US crop was confirmed to be contaminated with unapproved engineered genes, possibly due to accidental cross-pollination with conventional crops. The US government has since declared the rice safe for human consumption, and exports to some countries have since resumed.

3

Development of Aptamers

The detection of environmentally harmful substances by biosensors is a fast, easy to handle and cost-effective method. A lot of developments in the area of biosensors are described (Rogers, 2006), but for some substances the design of biosensors is difficult, particularly for such environmental analytes with high toxicity. Biosensors use biological recognition elements for the detection of the analyte and physical or electrochemical transducers for the transformation of the recognition reaction into an electrical signal. Most of the biological recognition elements used (e.g. enzymes, antibodies, cells and others) are sensitive to toxic substances.

Therefore new and robust receptors for environmental biosensors are needed. Aptamers could serve as alternative receptors for biosensors with high stability. Aptamers are single-stranded DNA or RNA oligonucleotides, which bind to their target with high selectivity and sensitivity because of their three-dimensional shape. The affinity reaction between the aptamer—fixed on a sensor surface—and its target can be measured, e.g. as a change of mass, thickness or optical properties of the receptor layer.

Concept of Aptamers

The term aptamer is derived from the latin word "aptus" which means fitting (Ellington & Szostak, 1990). Aptamers are short single stranded nucleic acid oligomers (ssDNA or RNA) with a specific and complex three-dimensional shape. They can well-fittingly bind to non-nucleic acid molecules comparable to antigen-antibody-binding.

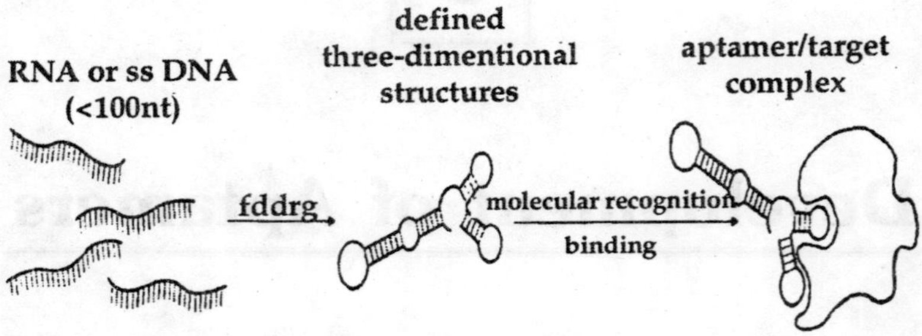

Fig.: Schematic representation of the functionality of aptamers

Binding of the aptamer to the target results from structure compatibility stacking of aromatic rings by electrostatic and van der Waals interactions, and hydrogen bondings, or from a combination of these effects (Hermann & Patel, 2000). The first aptamers developed consisted of unmodified RNA (Ellington & Szostak, 1990; Tuerk & Gold, 1990). In the following years single-stranded DNA aptamers for different targets were also described (Jayasena, 1999) as well as aptamers containing chemically modified nucleotides (Green *et al.*, 1995). Regarding to the binding behaviour no differences between DNA, RNA and chemically modified aptamers were found. Modified nucleic acids exhibit enhanced stability (Gold *et al.*, 1995). Aptamers were selected for a wide variety of different classes of targets. They range from small molecules, like nucleotides, cofactors or amino acids, over peptides, polysaccharides and proteins to complex structures like whole cells, viruses and single cell organisms. Aptamers can be developed both for nucleic acid binding molecules like polymerases (Tuerk & Gold, 1990) and also for molecules like growth factors (Jellinek *et al.*,

1993), antibodies (Tsai *et al.*, 1992), or hormones (Lin *et al.*, 1996), naturally not associated with nucleic acids.

Examples for targets and respective aptamers developed with specification of the affinity constants K_D

target group	*selection target*	*aplamer*	K_D *[nmol/L]*	*reference*
metallic ions	nickel	RNA	1000	(Hofmann *et al.* 1997)
nucleotides	flavin mononucleotide (FMN)	RNA	500	(Burgstaller&Famulok. 1994)
	adenosine triphosphate (ATP)	DNA	6000	(Huizenga & Szostak.1995)
amino acids	argininc	DNA	125000	(Harada&Frankel.1995)
	citrullin	RNA	65000	(Famulok, 1994)
amino glycosides	tobramycin	RNA	12	(Wang&Rando, 1995)
	neomycin	RNA	115	(Wallis *et al.*, 1995)
peptides	rev-peptid	RNA	20	(Xu & Ellington, 1996)
	vasopressin	L-DNA	900	(Williams *et al.*. 1997)
proteines	HIV 1-Rev	RNA	1	(Giver *et al.*. 1993)
	human thrombin	DNA	25	(Bock *et al.*, 1992)
	receptor tyrosine kinase (RET)	2'mod RNA	35	(Cerchia *et al.*,2005)
	vascular endothclial grow factor (VEGF)	2'mod RNA	0,14	(Green *et al.*, 1995)
	monoclonal antibody anti-MUCl	DNA	200	(Missailidis *el al*, 2005)
monosaccharides	galactose	RNA	10	(Kawakami *et al.*, 1998)
polysaccharides	dextran (sephadex)	RNA	n.b.	(Srisawat *et al.*, 2001)
	cellulose	DNA	300	(*Yang et al.*, 1998)
complex structures	trypanosomes	RNA	60	(Homann *&* Goringer, 1999)
	rous sarcoma virus	2'mod RNA	>40	(Pan *et al.*, 1995)
	anthrax-spores	DNA	n.b.	(Bruno &Kiel, 1999)
	brain cancer-endothelia) cells	DNA	n.b.	(Blank *et al.*, 2001)

Properties of Aptamers

Aptamer Structure: The three-dimensional shape arises on the one hand from intramolecular base pairings according to Watson and Crick. On the other hand "Non-Watson-Crick" base pairs or base stacking, pseudo-knots, "kissing"-complexes, parallel or anti-parallel triplexes and guanine-quadruplexes occur (Gilbert & Feigon, 1999; Li & Breaker, 1999).

The aptamers seem to bind to their target by "induced-fit" mechanism. That means that the aptamer changes its shape in the presence of the target, so that it will be able to bind. Possibly the target diversifies its shape in some respects to fit to the aptamer (Williamson, 2000). Another theory describes aptamers as very stable in their three-dimensional shape. Bivalent cations with the ability to link guanines and phosphate groups are incorporated. In this way the aptamer gets a high molecular rigidity and is able to recognise the target with high accuracy as well as to bind with high affinity and specificity (e.g. Blank, 2002).

Aptamer Affinity and Selectivity

For the detection and quantification of molecules in natural sciences and medicine more sensitive detection methods are needed, which are connected with the demand of more selective and sensitive recognition elements. Aptamers can be selected with high affinities and specificities for their targets comparable to those observed for antibodies. The dissociation constants (K_D) of most of the described aptamers are in the nanomolare to picomolare range (Famulok & Mayer, 1999; Tombelli *et al.*, 2005). But there is no correlation between the affinity of an aptamer and the nature of the target. Some aspects seem to contribute to a high affinity of aptamers like the ability to fold into an exceptionally stable three-dimensional structure or an interaction between the aptamer and the target over an extended domain of both binding partners. Chemical modifications of aptamer sequences could also improve the affinity (Klussmann, 2006).

Aptamers are also able to recognise their targets with high specificity. The selection of aptamers for a distinct epitope of a target enables the differentiation between closely related targets (e.g. differentiation between theophylline and caffeine; Jenison *et al.*, 1994). The distinction between chiral molecules is also possible (Michaud *et al.*, 2003). The selection process of specific aptamers out of a pool of aptamers can be modified by introducing negative selection steps or counter selection steps to direct the selection to aptamers for a specific subdomain of a target molecule. Aptamers can be selected for a wide variety of targets. However, there are some general prerequisites for a potential target. Functional groups should be present to form hydrogen bondings. Positively charged molecules cause electrostatic interactions with aptamers because of their negatively charged backbone. The aptamer selection for hydrophobic or negatively charged molecules (e.g. by phosphate groups) is more difficult.

In contrast to the classical production of antibodies by immunisation of animals or cell cultures, aptamer selection requires neither of them. Aside from ethical reasons, aptamer selection is advantageous for targets that are toxic and not immunogene. Moreover, it is possible to adapt aptamers during their *in vitro* selection to their application regarding the physicochemical conditions. It is not possible to influence these parameters in classical production of antibodies in hybridoma cell lines (Jayasena, 1999).

Aptamer Application

In spite of the very encouraging promises, the research on aptamers is still at the beginning. One disadvantage is the unavailability of a standardised protocol for aptamer development which is applicable without specific modifications for different targets. But aptamer research is catching up for faster results. Medical and pharmaceutical basic research as well as clinical diagnostic and therapy (e.g. inhibition of enzyme activities, blocking of receptor binding sites) offer a big field of application (Klug *et al.*, 1999; Hermann & Patel, 2000; Thiel, 2004). The big market of immunological diagnostics in the medical area will create an increasing competition between antibody and aptamer-based test kits. Some working groups are already starting to transfer antibody based diagnostic platforms to aptamers (Blank, 2002). The ELONA (Enzyme Linked OligoNucleotide Assay), based on ELISA technology, is one example for using aptamers instead of antibodies (Drolet *et al.*, 1996). In all probability, aptamers will be applied in standardised diagnostic test kits in the near future.

In the field of therapy are several aptamers for medically relevant targets, which will be tested regarding their application potential. The first approved aptamer with therapeutic function is the anti-human VEGF aptamer (VEGF: vascular endothelial grow factor) (Tucker *et al.,* 1999). Pegaptanib sodium as free acid of the oligonucleotide is used as the medicinal active component of the newly developed drug for the treatment of wet age related macular degeneration. The pharmaceutical product Macugen (pegaptanib sodium injection) from Pfizer Ltd./OSI Pharmaceuticals was approved in December 2004 (USA) and January of 2006 (Europe). Besides their use in the pharmaceutical area, aptamers are used for molecular detection in analytical systems (Clark & Remcho, 2002). Examples for the application of aptamers as specific ligands in affinity chromatography (Deng *et al.,* 2001), affinity capillary electrophoresis (German *et al.,* 1998; Connor & Mc Gown, 2005), capillary electrochromatography (Kotia *et al.,* 2000) and flow cytometry (Davis *et al.,* 1998) were described. In future, aptamers will also play an important role as new receptors in biosensors (O'Sullivan, 2002). Examples of aptamer applications mostly with optical detection of the binding reaction are described. E.g. a thrombin aptamer and an L-adenosin aptamer were used as biosensor receptors (Kleinjung *et al.,* 1998; Lee & Walt, 2000; Potyrailo *et al.,* 1998).

A further biosensor development is based on anti-human IgE aptamers and a quartz crystal microbalance for detection (Liss *et al.,* 2002). Aptamers have some advantages, which are very important for their use as molecular recognition element in analytical systems. Aptamers can be produced with very high accuracy and reproducibility by chemical synthesis, once their sequences are analysed. No variation between different production charges is anticipated (Kawazoe *et al.,* 1996). They are purified by stringent, denaturing conditions ensuring a very high purity. Reporter and adapter molecules like fluorophores and biotin may be introduced during synthesis at defined position of the aptamer sequence. Because the target recognition often results from the binding of relatively simple secondary or tertiary structure motifs, a subsequent modification of the aptamers regarding length or temperature stability is possible (Green *et al.*, 1995; Eaton *et al.*, 1997). The denaturation of aptamers and the binding of aptamer and target are reversible. The latter is very important for the use of aptamers in biosensors.

Development of Aptamers by in Vitro Selection

Aptamers are developed by use of an *in vitro* selection and amplification technology, called SELEX (Systematic Evolution of Ligands by Exponential Enrichment), or *in vitro* selection, or *in vitro* evolution. It was developed at the same time and independently from each other by J.W. Szostak (Ellington & Szostak, 1990) and L. Gold (Tuerk & Gold, 1990). Both approaches are based on the principle of the selective recognition of targets by the three-dimensional shape of single stranded nucleic acids. The basic principle of this *"in vitro* evolution" is the theorem of Darwin:

generation of a diversity of sequences → mutation,

selection of distinct phenotypical

characteristics selection, → amplification of the selected molecules → amplification.

Starting point of the SELEX process is a combinatorial library of single stranded nucleic acid molecules (RNA, ssDNA or modified nucleic acid molecules). Normally, this library exhibits 10^{15} different molecules (Gold *et al.*, 1995). This number exceeds the variants of the immune system by a factor of 10^7 (Janeway, 1993) and the variability of typical chemical libraries by a factor of 10^{10} (Eaton *et al.*, 1997). Therefore, for virtually every target, it should be feasible to select at least one possible binding partner (aptamer).

To exactly enrich these fitting structures out of the oligonucleotide library, the selection process takes place in presence of the target molecules. In doing so, target-binding oligonucleotides are separated from non-binding ones. The oligonucleotides interacting with the target will subsequently be amplified and used again for a binding reaction with the target in the following selection cycle. Because of the complexity of the library, the enrichment of aptamers by one single step of selection and amplification is not possible. Several repetitive selection cycles are needed, until target-specific aptamers dominate within the selected oligonucleotide pool. After cloning, individual aptamers of this pool will be sequenced and characterised subsequently (Klug & Famulok, 1994).

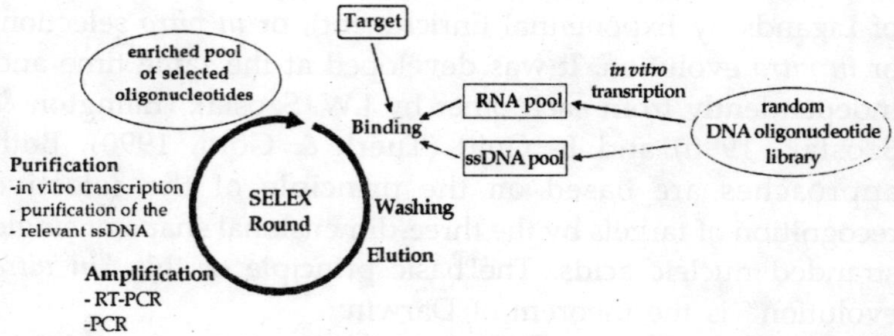

Fig.: General schematic representation of a SELEX process for selecting RNA or DNA aptamers. Starting point of each SELEX process is a synthetic random DNA oligonucleotide library consisting of a multitude of ssDNA fragments (approximately 10^{15}) with different sequences. This library can be used directly for the selection of DNA aptamers. For the selection of RNA aptamers the library has to be transferred into RNA by in vitro transcription. The aptamer selection is characterised by the repetition of successive steps consisting of binding between oligonucleotides and target, removal of unbound oligonucleotides by washing of the binding complex, elution of target-bound oligonucleotides followed by their amplification and purification. A new enriched pool of selected oligonucleotides is generated and used for the next selection round. In general, 6 to 20 SELEX rounds are needed for the selection of high affine, target-specific aptamers. Having finished the last round, the enriched aptamer pool is cloned and several individual aptamers have to be characterised.

If one compares RNA- and DNA-SELEX procedures, both show advantages and disadvantages. The crucial difference lies in the additional *in vitro* transcription step for the RNA-SELEX to transfer the initial synthetic DNA oligonucleotide library into an RNA-library, whereas for the DNA-SELEX the library can be used directly. The selected RNA-oligonucleotides of every RNA-SELEX round has to be reverse transcribed into cDNA and subsequently amplified by PCR. The new RNA pool for the next selection round again is generated by *in vitro* transcription from the resulting PCR products. During the DNA-SELEX procedure, the generation of the new selected DNA oligonucleotide pool for the next round is carried out by purification of the relevant single strands from the double stranded PCR products. Besides of these differences during the process, RNA and DNA aptamers differ primarily in their resistance to enzymatic digestion and their potential for modifications. Natural RNA has a smaller stability to endonucleases, e.g. in biological fluids than DNA. Modified RNA (e.g. with 2'-Fluoro-, 2'-Amino- or 2'-O-Methyl-modifications) and DNA (L-isomers) are more stable in this regard (Gold *et al.*, 1995; Williams *et al.*, 1997; James, 2000). One important advantage of DNA compared to RNA is the possibility to directly introduce reporter molecules with different functions during PCR. These molecules can comprise reporter compounds like fluorophores for detection, functional groups and molecules for immobilisation or for single strand partitioning (e.g. amino-, phosphate-, thiol-groups, spacers, biotin). During the RNA-SELEX the incorporation of such molecules at defined positions is more difficult or even not possible (Conrad & Ellington, 1996).

The direct interaction of the oligonucleotides with the target during the binding steps of a SELEX process is closely application-oriented. An in-depth knowledge about the respective target is not necessary for the aptamer selection, which is a big advantage. Usually single, well defined targets, like peptides, proteins or antibiotics, are used for SELEX. However, aptamer selection is also feasible for complex or unknown structures, like whole cells.

One of the crucial steps of a SELEX process with outstanding importance for the selection of aptamers with high affinity and specificity is the efficient separation between target-binding and non-binding oligonucleotides. The use of affinity chromatography with immobilisation of target on column material like sepharose or agarose is a conventional method for the separation step (Tombelli *et al.*, 2005). For this purpose, substantial amounts of target are necessary to obtain a high efficient loading of the column. Another method of separation is the ultrafiltration by use of nitrocellulose filters with distinct molecular weight cut-offs. Thereby, losses of target-binding oligonucleotides as well as unspecific interactions of oligonucleotides with the membrane resulting in their unspecific enrichment are possible but undesired effects.

A further possible method of separation constitutes the use of magnetic beads. These are paramagnetic polystyrene beads with a ferric content of 12-15 per cent. In contrast to the aforementioned methods this method requires only very small amounts of target and enables a very simple handling. Magnetic separation technology is fast, simple and effective. The separation of target-bound and unbound oligonucleotides is efficient and stringent washing steps are possible (Stoltenburg *et al.*, 2005). Moreover, the magnetic beads allow parallel procedures to some degree, "by hand" as well as automated. Magnetic beads are available with different functional groups (e.g. epoxy-, tosyl-, biotin-, amino-, carboxyl-, or antibody layers). Therefore, various targets can be immobilised like peptides, proteins, antigens, antibodies as well as nucleic acids, carbohydrates and other molecules (Safarik & Safarikova, 2004).

Selex-variants

The SELEX procedure allows the adaptation of the conditions to further applications of the aptamers already during the process. Numberless variants of the originally by Tuerk & Gold (1990) established procedure were described. A short overview about the mostly used types of SELEX will be given here:

- The *Counter-SELEX* generates aptamers which are able to discriminate between closely related structures. A selection step to the related target is introduced and eliminates aptamers from the oligonucleotide pool, which are not able to discriminate between the related structures. An impressive example is the theophylline aptamer. The SELEX process was configured by Jenison *et al.* (1994) in such a way, that the selected nucleic acid ligands bind their target theophylline 1000times better than the related structure caffein, which differs only by a methyl group at the nitrogen N-7 of the purine ring.

- A *Negative-SELEX* is used to minimise the co-selection of unwanted nucleic acid ligands. Previous to the binding reaction with the target, the nucleic acid ligands will be incubated with molecules, which should not be recognised. The oligonucleotides binding to the unwanted structures, will be removed from the pool. An example was shown by Blank (2002), who avoided in a SELEX for pathological mutated endothelial cells any background binding by unspecific attachment to cell surfaces, binding to the matrix as well as the enrichment of endothelial unspecific aptamers.
- In a *Blended SELEX* the nucleic acid molecules will be enlarged by non-nucleic acid molecules like peptides that give additional properties to the aptamers beyond the binding capability. For instance, aptamers that have the power to de-activate the enzymes they recognise were selected by this method (Smith *et al.*, 1995).
- Within a *Chimeric SELEX* two or more different libraries will be used for the production of chimeric aptamers with more than one wanted feature or which are able to function in different ways. Each of the parent libraries will be selected first to a distinct feature. The selected aptamers are fused subsequently in such a way, that they are getting several properties (Burke & Willis, 1998).
- The *Deconvolution-SELEX* is appropriate to find aptamers binding complex targets. With this procedure the discrimination between relevant aptamers (binding to distinct target structures within the complex mixture) and irrelevant oligonucleotides within the selected pool is possible (Blank *et al.*, 2001).

- The *Photo-SELEX* is another variant yet with the result of aptamers bearing photo-reactive groups. These aptamers are able to bind and to photo cross-link to a target and/or to photoactivate a target molecule (Sampson, 2003). Golden *et al.* (2000) developed a SELEX for the identification of nucleic acid binding proteins. The aptamers containing bromo-uracil will be coupled covalently with the protein by UV irradiation. This highly stable covalent binding enables stringent washing steps and the execution of high-throughput SELEX processes.
- *SPIEGELMER-Technology* is a completely different method. Ordinary D-nucleic acids will be selected for the mirrored (enantiomer) target. The aptamers resulting will be synthesised as L-isomers and bind now to the originally unmirrored target. Spiegelmers are of high significance for medical application, because they will not be degraded by human enzymes (Klussmann *et al.,* 1996; Leva *et al.,* 2002).

FluMag-SELEX

The FluMag-SELEX (Stoltenburg *et al.,* 2005) is a special method for the selection of DNA aptamers.

This procedure is characterised by the use of fluorescence labelled ssDNA molecules for quantification and magnetic beads for target immobilisation.

For the assessment of the SELEX procedure the quantification of the enriched target-binding oligonucleotides as well as the amount of unbinding oligonucleotides of each selection round have to be determined. Mostly radioactive markers will be used for the quantification during a SELEX process. Although this is a very sensitive method, enabling the detection of slightest amounts of nucleic acids, the drawback of this technology is the necessity to manage the whole process in an isotope laboratory, which is very cost-intensive. Moreover, the use of radioactive material is not environmentally compatible and holds a health risk for employees. Alternatively, fluorescence labels may be used for quantification.

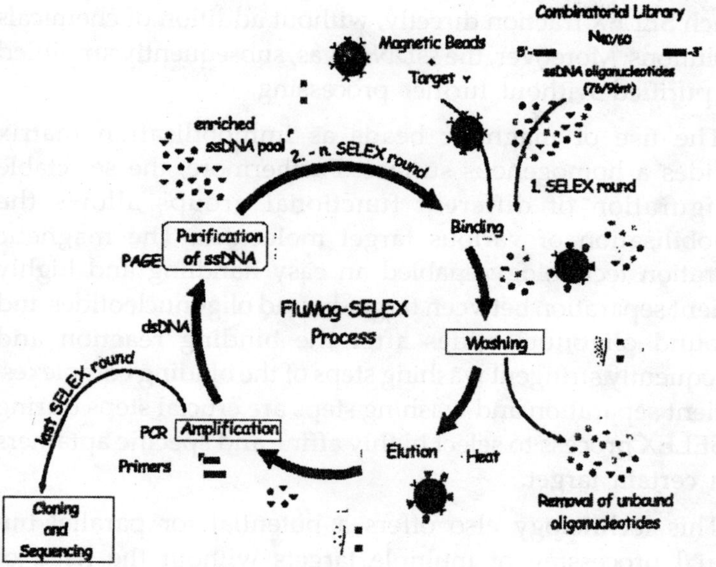

Fig.: Schematic representation of the fluorescence-monitored in vitro *selection procedure, the so-called FluMag-SELEX, for selecting of DNA aptamers for specific target molecules immobilised on magnetic beads. The starting ssDNA combinatorial library consisting of oligonucleotides with a centrally randomised region of 60 nucleotides, flanked by two specific regions, was incubated with target beads (target-coated magnetic beads, TMB) for binding. Unbound oligonucleotides were washed away and bound oligonucleotides were eluted from target beads by heat treatment. The selected oligonucleotides were amplified by PCR using special modified primers. The relevant ssDNA was subsequently purified from the PCR products. The resulting enriched DNA pool was used in the next SELEX round for binding with target beads. By labelling of selected oligonucleotides after the first SELEX round with fluorescein (F), their quantification in further rounds and thus monitoring of the enrichment of target-specific aptamers was enabled. During the last round the selected aptamer pool was cloned and several individual aptamers were characterised.*

By this means, the operation in a molecular biological standard laboratory is possible, additional costs are avoided and a satisfactory high sensitivity for quantification and detection of nucleic acids is gained. After the first round of the

FluMag-SELEX process the selected DNA was labelled with fluorescein by PCR using a 5'-modified primer. Thus, it was possible to quantify the ssDNA during further selection rounds in each SELEX fraction directly, without addition of chemicals or solutions. Moreover, the ssDNA was, subsequently amplified and purified without further processing.

The use of magnetic beads as immobilisation matrix provides a homogenous surface. Furthermore, the selectable configuration of different functional groups allows the immobilisation of various target molecules. The magnetic separation technology enabled an easy handling and highly efficient separation between target-bound oligonucleotides and unbound oligonucleotides after the binding reaction and subsequently stringent washing steps of the binding complexes. Efficient separation and washing steps are crucial steps during the SELEX process to select highly affine and specific aptamers for a certain target.

This technology also offers a potential for parallel but manual processing of multiple targets without the need of expensive robotics. Furthermore, additional steps can be introduced in the aptamer selection process. Negative selection steps should be performed before the binding reaction with the target in each selection round and are necessary to remove unspecific binding oligonucleotides or to minimise the risk of their enrichment during the SELEX process. Moreover, negative selection steps against related target molecules can be used to enhance the specificity of selected aptamers or to direct the selection process to certain epitopes within the target molecules.

The FluMag-SELEX is a universal aptamer selection procedure and adaptable to a wide variety of target molecules. After finishing of the SELEX process the selected aptamers undergo a further characterisation by sequence analysis and binding studies. The combination of target-coated magnetic beads and fluorescence-labelled aptamers can also be used in different binding assays for rapid and comparative determination of relative binding capabilities of individual aptamers generated in a SELEX experiment. This technology enables also the determination of affinity constants of the selected aptamers and investigations to their cross reactivities by affinity elution assays.

DNA Aptamers Selected by Flumag-SELEX

The *in vitro* aptamer selection procedure, FluMag-SELEX, based on magnetic beads as immobilisation matrix for target molecules and utilising fluorescence labelling for quantification of DNA during the SELEX process was used to select DNA aptamers for different classes of targets. The range of targets reached from small organic molecules over peptides and proteins to complex targets.

DNA Aptamers for Streptavidin

Streptavidin is a tetrameric protein which binds very tightly to the small molecule biotin ($K_D \sim 10^{-15}$ mol/L). This interaction is one of the strongest biological and non-covalent interactions known. Therefore, the streptavidin/biotin system is widely used in scientific research, e.g. as universal tools in different test systems for immunology or molecular diagnostic (to immobilise molecules onto surfaces or to interlink molecules).

Streptavidin as a well known protein was immobilised on tosylactivated magnetic beads (Dynabeads, Invitrogen/ Dynal, UK) and used as a model target to establish and to optimise the FluMag-SELEX process (Stoltenburg *et. al.*, 2005). The aptamer selection process was started with a random oligonucleotide library, in which each oligonucleotide consisted of a central random region of 60 nt flanked by different specific sequences of 18 nt in each direction. After the binding reaction and several washing steps of the binding complex the oligonucleotides which were bound to the streptavidin-coated magnetic beads, were eluted by heat treatment in each selection round and quantified by fluorescence measurement. Thus, from the second round the enrichment of target-binding aptamers was monitored during the selection process. An increasing amount of oligonucleotides bound to streptavidin-coated magnetic beads was observed after rounds 6, 7, and 8.

In further rounds the amount of eluted oligonucleotides remained nearly constant, possibly because of full occupation of binding sites on the beads. During the last three selection rounds a negative selection step was introduced against tosylactivated magnetic beads, which were used as a matrix for the immobilisation of streptavidin. Enrichment of unspecific binding oligonucleotides to the bead matrix was not detected. After round 13 the selection process was finished and the selected aptamer pool was cloned and further characterised. Five groups of sequences were identified containing a different number of aptamer clones with mostly identical sequences within each group.

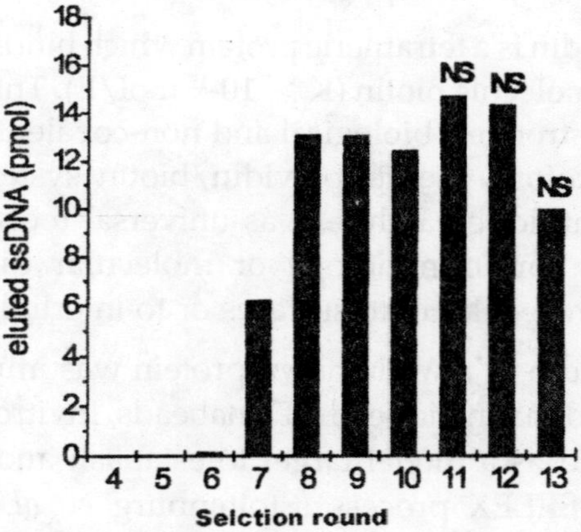

Fig.: Enrichment of streptavidin-specific aptamers during the FluMag-SELEX process. The bar graph shows the amounts of ssDNA eluted from streptavidin-coated magnetic beads in each selection round. In rounds 11, 12, and 13 a negative selection step (NS) was introduced to remove ssDNA unspecifically bound to the bead matrix

One representative from each group, aptamer clones 2, 26, 30, 31 and 33, was used for further characterisations in individual binding assays. These assays were performed according to the FluMag-SELEX conditions using streptavidin-coated magnetic beads as target and one of the selected and fluorescein-labelled aptamers. They enable a rapid and comparative determination of the relative binding capabilities of different aptamers generated in a SELEX experiment. The best binding results were obtained with aptamers 26 and 31.

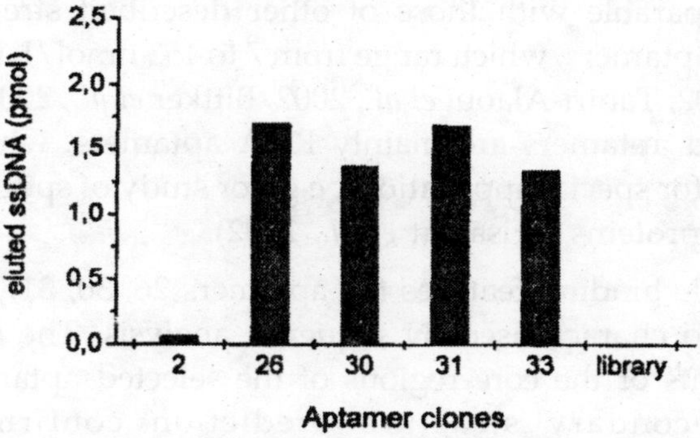

Fig.: Binding assays with the individual aptamer clones 2, 26, 30, 31 and 33, to test their ability to bind to streptavidin-coated magnetic beads. Each assay was performed according to the FluMag-SELEX conditions (streptavidin-coated magnetic beads, fluorescence-labelled aptamer). The unselected random DNA oligonucleotide library was used as the negative control. The bar graph shows the amounts of aptamer eluted from streptavidin-coated magnetic beads after the binding reaction

The binding capability for the aptamers 30 and 33 was slightly lower. In contrast, aptamer 2 revealed only a very low binding capability to streptavidin-coated magnetic beads. The unselected random oligonucleotide library was used as the negative control in these assays, and showed no binding to target beads.

To determine the affinities of individual aptamers 26, 30, 31, and 33 to their target, the dissociation constants (K_D values) were ascertained. For this purpose binding reactions with a constant number of streptavidin-coated magnetic beads (1×10^7) as target and concentration series of each fluorescein-labelled aptamer (0 - 188 n mol/L) were performed. Saturation curves for each aptamer were obtained, and the dissociation constants (K_D) were calculated by non-linear regression analysis. The lowest K_D of 56.7 ± 8.2 nmol/L was obtained for the aptamer 31. The K_D values of the other three aptamers were slightly higher: 59.0 ± 7.5 nmol/L for the aptamer 33, 65.5 ±

8.6 nmol/L for the aptamer 30, and 85.2 ± 5.3 nmol/L for the aptamer 26.

These dissociation constants in the low nanomolar range are comparable with those of other described streptavidin-specific aptamers, which range from 7 to 153 nmol/L (Srisawat *et al.*, 2002, Tahiri-Alaoui *et al.*, 2002, Bittker *et al.*, 2002). These described aptamers are mainly RNA aptamers, which were selected for special applications, e.g. for study of special RNAs and riboproteins (Srisawat *et al.*, 2002).

Beside binding features the aptamers 26, 30, 31, 33 and 2 were also characterised by sequence analysis. The sequence alignments of the core regions of the selected aptamers and their secondary structure prediction confirmed the correspondences and differences among the five aptamers derived from the individual binding assays. The aptamers 26 and 31 contained a consensus sequence of 17 nucleotides near the 3'-end of the core region. This conserved motif formed a stem-loop structure in which the stem consisted of three guanosine-cytidine base pairs.

The sequences of the aptamers 30 and 33 were completely different from those of the aptamers 26 and 31. There was a thymidine-rich sequence at the 3'-end of the core region with short homologous sections. The secondary structure analysis also predicted the formation of a stem-loop in this motif. The stem consisted of two guanosine-cytidine base pairs and the loop was characterised by five or six successive thymidines. It is conceivable that both stem-loop structures play an important role during the process of target recognition and binding of the four aptamers to streptavidin-coated magnetic beads. The stems could have a stabilising function whereas the bases in the loops could be responsible for binding. In contrast, none of the identified sequence motifs was found in aptamer 2.

Further experiments were performed with the aptamers 26, 30, 31 and 33 to investigate the effect of biotin on the interaction of these aptamers with streptavidin, because biotin is the natural binding partner of streptavidin with an extraordinary high affinity (~10^{15} mol/L). For this purpose the binding complexes consisting of aptamer saturated streptavidin-coated magnetic beads were incubated in a biotin solution. All four aptamer clones were released from the binding complexes by biotin in a concentration dependent manner.

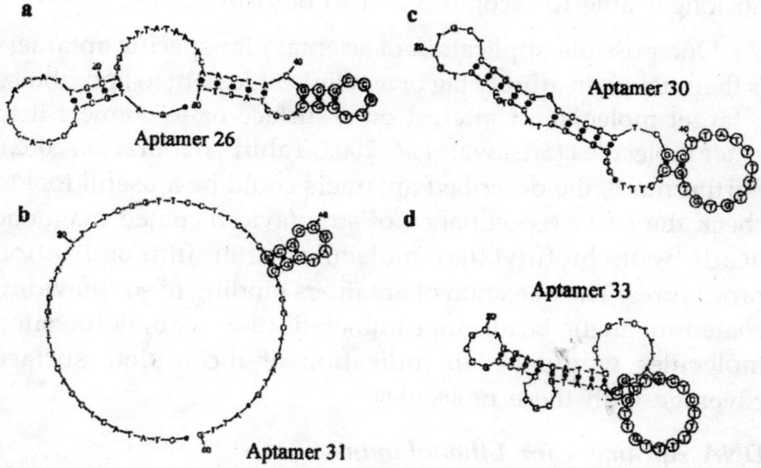

***Fig.:** Secondary structure prediction for the core regions of the aptamers 26 (a), 31 (b), 30 (c), and 33 (d). The stem-loop structures in the conserved regions of these aptamers are marked by circles*

The strongest effect was found for the aptamers 30 and 33. Both aptamers were completely eluted from target beads by 600 - 2500 nmol/L biotin, a concentration, which is equimolar with or higher than that of biotin-binding sites of the streptavidin-coated magnetic beads (1×10^7) used in these assays. In contrast, biotin had less effect on binding of the aptamers 26 and 31 to the target beads. Both of these aptamers were released much more slowly. Complete elution in the presence of 2500 nmol/L biotin was not observed. Furthermore the binding of the four aptamers to streptavidin-coated magnetic beads was blocked, if the beads were pre-incubated with sufficient biotin to saturate all biotin-binding sites. The separation or blocking effect of biotin on the aptamer binding is due to their competition to bind to streptavidin. It is possible that the aptamer-binding site is near to or even at the biotin-binding site of streptavidin. On the other hand, the described aptamers were selected for unmodified streptavidin-coated magnetic beads. The binding of biotin could cause a conformational change of streptavidin so that the aptamers are no longer able to recognise and to bind to it.

One possible application of streptavidin-specific aptamers is their use as an affinity tag or as a linker element to immobilise a target molecule of interest on a surface or to connect it to other molecules (Srisawat *et al.*, 2002, Tahiri-Alaoui *et al.*, 2002). Furthermore, the described aptamers could be a useful tool to check the surface occupancy of streptavidin-coated magnetic beads with biotinylated molecules after immobilisation procedures. The detection of aptamers binding to streptavidin-coated magnetic beads after immobilisation with biotinylated molecules would be an indication of incomplete surface coverage with these molecules.

DNA Aptamers for Ethanolamine

Ethanolamine is a very small organic molecule ($M_w = 61.08$) with biological, medical, and industrial relevance. It is involved in the biosynthesis of acetylcholine, known as a component in phospholipids, and associated with the storage disease ethanolaminosis. Moreover, it is used for scrubbing certain acidic gases and is also used in surface active agents, cleaners, emulsifiers or cosmetic products. Ethanolamine seems to be the smallest aptamer target so far. It is characterised by a short carbon chain (2 C) and two functional groups (amino group, hydroxyl group).

Ethanolamine binding DNA aptamers were selected from an initial random oligonucleotide library (96mers), which was also used for the selection of streptavidin binding aptamers (Mann *et al.*, 2005). The target ethanolamine was immobilised on epoxy functionalised or tosylactivated magnetic beads (Dynabeads, Invitrogen/Dynal, UK). After 10 selection rounds the FluMag-SELEX process was finished and the selected aptamer pool was cloned. 37 aptamer clones from the selected pool were analysed. There were two main groups (I and II) with 21 and 11 homologous aptamer sequences, respectively. A third group (III) was represented only by one aptamer sequence. Sequence alignments revealed in all three groups a conserved guanosine-rich section of 16 nucleotides at the 5'-end of the core region of the aptamers. The secondary structure of one representative aptamer (EA#14.3, EA#9.4) of the two main groups is characterised by the formation of three stem-loop structures predicted by a sequence folding programme (mfold) (Zuker, 2003).

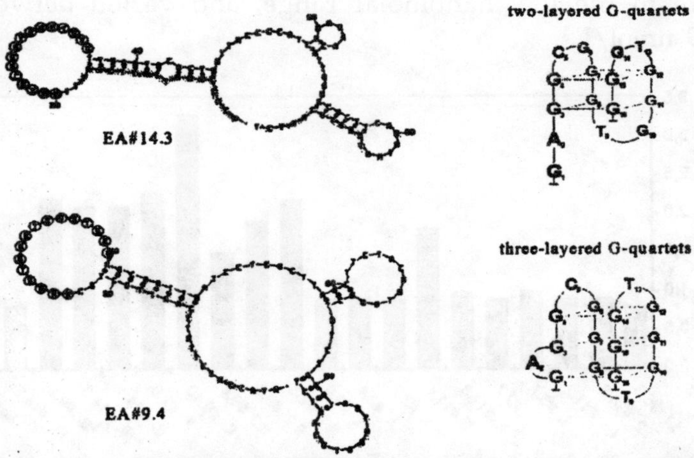

Fig.: Secondary structure prediction for the ethanolamine binding aptamers #14.3 (a) and #9.4 (b). The primer regions are highlighted in green and the G-rich consensus sequence of 16 nucleotides is marked by circles. Based on the nucleotide composition the consensus sequence could fold into a two- or three-layered G-quartet structure (c, d)

It was also found, that the G-rich consensus sequence is located in the single stranded region of one of the three stem-loop structures. Therefore, it seems possible that three-dimensional folding of the consensus sequence leads to a two- or three layered G-quadruplex structure, described for some other aptamers containing a similar G-rich sequence, e.g. the thrombin aptamer (Bock *et al.*, 1992).

The selected aptamers for ethanolamine were also tested in individual binding assays for their ability to bind to ethanolamine immobilised on magnetic beads, according to FluMag-SELEX conditions.

Differences in the binding capability between the three aptamer groups were observed. The average amount of bead-bound aptamers was in group II about 50 per cent higher than in group I and III. The random oligonucleotide library used as negative control was not able to bind to the ethanolamine-coated magnetic beads. Both tested representative aptamers of the two main groups EA#14.3 and EA#9.4 were able to bind their target with high affinity. The dissociation constants (K_D) were in the low nanomolar range, and varied between 6-19 nmol/L^{-1}.

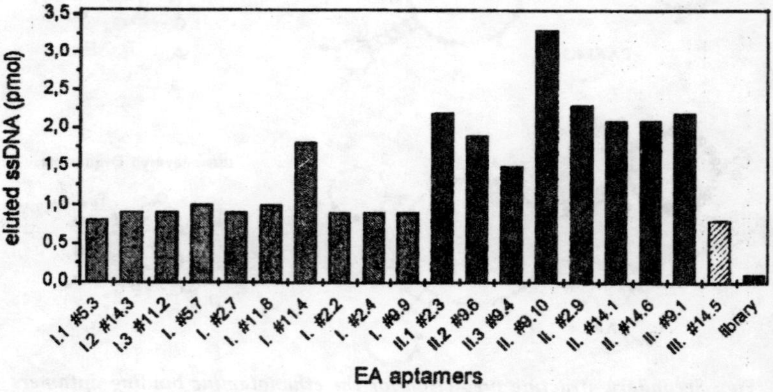

Fig.: Comparison of binding abilities of all aptamer clones of group I, II and III. The binding ability of all individual aptamers was tested in single binding assays with ethanolamine-coated magnetic beads as target. The bar graph shows the amount of aptamers eluted from target beads after the binding reaction of each assay. The unselected random DNA oligonucleotide library served as negative control

Because of the simple structure of ethanolamine, which is present in many other organic molecules, the specificity of the selected aptamers EA#14.3 and EA#9.4 was investigated by affinity elution assays. In these assays the binding complexes consisting of ethanolamine-coated magnetic beads and bound aptamers were incubated in a test solution. The test solution contained an excess of ethanolamine or another potential target molecule. The organic molecules in the test solution competed against ethanolamine-coated magnetic beads as target for the aptamers. Only if cross reactivity exists the bead-bound aptamers are released from the binding complexes by binding to the organic molecules in the test solution. The results indicate that the aptamers are also able to bind to free ethanolamine in solution and to related molecules like the ethanolamine derivatives di- and triethanolamine.

They were not able to bind to structurally and functionally similar molecules like serine, choline or glycine. The applications of the ethanolamine binding aptamers are multifarious. More complex molecules which contain an accessible, ethanolamine related structure can be detected by these aptamers or can be purified from a complex sample. The aptamers could be also used as an affinity tag or as a linker element to immobilise a target molecule of interest on a surface.

4

Anaerobic Digestion

Anaerobic digestion (AD) is the harnessed and contained, naturally occurring process of anaerobic decomposition. An anaerobic digester is an industrial system that harnesses these natural process to treat waste, produce biogas that can be used to power electricity generators, provide heat and produce soil improving material.

Anaerobic digesters have been around for a long time and they are commonly used for sewage treatment or for managing animal waste. Increasing environmental pressures on waste disposal have increased the use of AD as a process for reducing waste volumes and generating useful by-products. It is a fairly simple process that can greatly reduce the amount of organic matter which might otherwise end up in landfills or waste incinerators.

Almost any organic material can be processed in this manner. This includes biodegradable waste materials such as waste paper, grass clippings, leftover food, sewage and animal waste. Anaerobic digesters can also be fed with specially grown energy crops to boost biodegradable content and hence increase biogas production. After sorting or screening to remove inorganic or hazardous materials such as metals and plastics, the material to be processed is often shredded, minced, or hydrocrushed to increase the surface area available to microbes in the digesters and hence increase the speed of digestion. The material is then fed into an airtight digester where the anaerobic treatment takes place.

Stages of Anaerobic Digestion

There are Two Conventional Operational Temperature Levels:

- *Mesophilic* which takes place optimally around 37-41°C or at ambient temperatures between 20-45°C with mesophile bacteria.
- *Thermophilic* which takes place optimally around 50-52°C at elevated temperatures up to 70°C with thermophile bacteria.

The residence time in a digester varies with the amount of feed material, type of material and the temperature. In the case of mesophilic digestion, residence time may be between 15 and 30 days. In the case of mesophilic UASB digestion hydraulic residence times (1hour-1day) and solid retention times (<90 days) are separated. In the thermophilic phase the process can be faster, requiring only about two weeks to complete.

Thermophilic digestion is more expensive, requires more energy and is less stable than the mesophilic process. Therefore, the mesophilic process is still widely in use.

Many continuous digesters have mechanical or hydraulic devices to mix the contents and to allow excess material to be continuously extracted to maintain a reasonably constant volume.

The digestion of the organic material involves a range of many different species of naturally occurring bacteria, all doing a different job at a different step in the digestion process. Maintaining suitable conditions in the digester is essential in maintaining a healthy bacterial population.

Four Stages of Anaerobic Digestion have been recognised:

1. The first is hydrolysis, where complex organic molecules are broken down into simple sugars, amino acids, and fatty acids with the addition of hydroxyl groups.
2. The second stage is acidogenesis, where a further breakdown into simpler molecules, i.e., volatile fatty acids (e.g., acetic, propionic, butyric, valeric) occurs, producing ammonia, carbon dioxide and hydrogen sulphide as by-products.
3. The third stage is acetogenesis, where the simple molecules from acidogenesis are further digested to produce carbon dioxide, hydrogen and mainly acetic acid.
4. The fourth stage is methanogenesis, where methane, carbon dioxide and water are produced.

By-products of Anaerobic Digestion

There are three principal by-products of anaerobic digestion:

- Biogas, a gaseous mixture comprising mostly of methane and carbon dioxide, but also containing a small amount hydrogen and occasionally trace levels of hydrogen sulphide. Biogas can be burned to produce electricity, usually with a reciprocating engine or microturbine. The gas is often used in a cogeneration arrangement, to generate electricity and use waste heat to warm the digesters or to heat buildings. Excess electricity can be sold to electricity suppliers. Electricity produced by anaerobic digesters is considered to be green energy and may attract subsidies such as Renewables Obligation Certificates.

Since the gas is not released directly into the atmosphere and the carbon dioxide comes from an organic source with a short carbon cycle biogas does not contribute to increasing atmospheric carbon dioxide concentrations; because of this, it is considered to be an environmentally friendly energy source. The production of biogas is not a steady stream; it is highest during the middle of the reaction. In the early stages of the reaction, little gas is produced because the number of bacteria is still small. Towards the end of the reaction, only the hardest to digest materials remain, leading to a decrease in the amount of biogas produced.

- The second by-product (acidogenic digestate) is a stable organic material comprised largely of lignin and chitin, but also of a variety of mineral components in a matrix of dead bacterial cells; some plastic may be present. This resembles domestic compost and can be used as compost or to make low-grade building products such as fibreboard.
- The third by-product is a liquid (methanogenic digestate) that is rich in nutrients and can be an excellent fertilizer dependent on the quality of the material being digested. If the digested materials include low levels of toxic heavy metals or synthetic organic materials such as pesticides or PCBs, the effect of digestion is to significantly concentrate such materials in the digester liquor. In such cases, further treatment will be required in order to dispose of this liquid properly. In extreme cases, the disposal costs and the environmental risks posed by such materials can offset any environmental gains provided by the use of biogas. This is a significant risk when treating sewage from industrialised catchments.

Nearly all digestion plants have ancillary processes to treat and manage all of the by-products. The gas stream is dried and sometimes sweetened before storage and use. The sludge liquor mixture has to be separated by one of a variety of ways, the most common of which is filtration. Excess water is also sometimes treated in sequencing batch reactors (SBR) for discharge into sewers or for irrigation.

Digestion can be either *wet* or *dry*. Dry digestion refers to mixtures which have a solid content of 30 per cent or greater, whereas wet digestion refers to mixtures of 15 per cent or less.

Reactor Types

There is a range of types of anaerobic digesters, however the two main types of operations are batch and continuous.

Batch is the simplest, with the biomass added to the reactor at the beginning and sealed for the duration of the process. Batch reactors can suffer from odour issues which can be a severe problem during emptying cycles.

In the continuous process, which is the more common type, organic matter is constantly added to the reactor and the end products are constantly removed, resulting in a much more constant production of biogas.

Considerations

To be economically viable, there must be a market for the end products. Biogas can be sold or used in almost all parts of the world, where it will offset demand on fossil fuel stocks. The digester liquor is suitable for use as a fertilizer, although frequently supplemental nutrients need to be added.

The sludge component, even when dried and available as a soil conditioner, is not easily disposed of. However, it has its uses in non-agricultural areas, such as golf courses, and as cover for landfills. In some localities, the sludge itself is used as a fuel in heating systems, and the residual ash is disposed of in a landfill.

Contribution to Prevention of Climate Change

Production of Renewable Fuel: Processing biodegradable waste using anaerobic digestion helps to reduce global warming. The carbon in biodegradable waste is part of a complete carbon-cycle: the carbon released from the combustion of biogas was removed by plants in the recent past, and does not contribute to the global accumulation of carbon in the same manner that fossil fuels do. Furthermore, if this waste was landfilled it would break down naturally and the biogas would escape directly into the atmosphere. Using the biogas for energy is an intermediate use that does not affect the overall cycle. In this way anaerobic digestion is considered to be a sustainable technology and biogas is considered to be a renewable fuel.

Associated Technologies

Mechanical Biological Treatment

New developments in anaerobic digestion have led to systems being integrated with sorting units. Mixed waste streams such as unsorted household waste can undergo a mechanical pretreatment stage. These systems come under the category of mechanical biological treatment. They enable the recovery of the organic fraction of the waste in a form that can be processed in anaerobic digesters.

Inhibition of Methanogenesis and Production of Alcohols

Anaerobic digestion can be inhibited from reaching the methanogenic stage. The organic acids (i.e., carboxylic acids) from the acidogenic and acetogenic stages of the digestion can be recovered. The acids can then undergo further chemical transformations into useful chemicals or fuels.

Potential in the Hydrogen Economy

As anaerobic digestion is a renewable source of methane it offers the potential to contribute to the hydrogen economy:

Steam methane reforming (SMR) is the most common method of producing commercial bulk hydrogen. It is also the least expensive method. At high temperatures (700-1100°C) and in the presence of a metal-based catalyst, steam reacts with methane to yield carbon monoxide and hydrogen.

$$CH_4 + H_2O \rightarrow CO + 3\ H_2$$

The United States produces nine million tons of hydrogen per year, mostly with steam reforming of natural gas. This process is different from catalytic reforming, an oil refinery process that also produces significant amounts of hydrogen along with high octane rating gasoline.

Acetogenesis

Acetogenesis is a process through which acetate is produced by anaerobic bacteria from a variety of energy (for example, hydrogen) and carbon (for example, carbon dioxide) sources. The different bacterial species that are capable of acetogenesis are collectively termed acetogens.

Chemical Equation

$$2\,CO_2\text{ (aq)} + 4H_2\text{ (aq)} = CH_3COOH\text{ (aq)} + 2H_2O$$

Acetogenesis and Anaerobic Digestion

Acetogenesis is the third stage of anaerobic digestion

- *Hydrolysis:* Particulates are solubilised and large polymers, converted into simpler monomers;
- *Acidogenesis:* Simple monomers are converted into volatile fatty acids;
- *Acetogenesis:* Volatile fatty acids are converted into acetic acid, carbon dioxide, and hydrogen; and
- *Methanogenesis:* Acetate are converted into methane and carbon dioxide, while hydrogen is consumed.

Anaerobic digestion is a complex biochemical process of biologically-mediated reactions by a consortia of microorganisms to convert organic compounds into methane and carbon dioxide. It is a stabilisation process, producing odour, pathogens, and mass reduction.

Hydrolytic bacteria form a variety of reduced end-products from the fermentation of a given substrate. One fundamental question which arises, concerns the metabolic features which control carbon and electron flow to a given reduced end-product during pure culture, and mixed methanogenic cultures of hydrolytic bacteria. *Thermoanaerobium brockii* is a representative thermophilic, hydrolytic bacterium, which ferments glucose, via the Embden-Meyerhof Parnas Pathway.

T. brockii is an atypical hetero-lactic acid bacterium because it forms molecular hydrogen (H_2), in addition to lactic acid and ethanol. The reduced end-products of glucose fermentation are enzymatically-formed from pyruvate, via the following mechanisms: lactate by fructose 1-6 all-phosphate (F_6P) activated lactate dehydrogenase; H_2 by pyruvate ferredoxin oxidoreductase and hydrogenase; and ethanol via NADH- and NADPH-linked alcohol dehydrogenase.

By its side, the acidogenic activity was found in the early 20th century, but it was not until mid-'60s that the engineering of phases separation was assumed in order to improve the stability and waste digesters treatment. In this phase, complex molecules (carbohydrates, lipids, proteins) are depolymerised into soluble compounds by hydrolytic enzymes (cellulases, hemicelulases, amylases, lipases and proteases). The hydrolysed compounds are fermented into volatile fatty acids (acetate, propionate, butyrate, and lactate), neutral compounds (ethanol, methanol), ammonia, hydrogen and carbon dioxide

Acetogenesis is one of the main reactions of this stage, in this, the intermediary metabolites produced are metabolised to acetate, hydrogen and carbonic gas by the three main groups of bacteria:

- Homoacetogens;
- Syntrophes; and
- Sulphoreductors.

For the acetic acid production are considered three kind of bacteria:

- *Clostridium aceticum;*
- *Acetobacter woodii*; and
- *Clostridium termoautotrophicum.*

Winter Wolfe, in 1979, demonstrated that *A. wodii* in syntrophic association with *Methanosarcina* produce methane and carbon dioxide from fructose, instead of three molecules of acetate. *C. thermoaceticum* and *C. formiaceticum* are able to reduce the carbonic gas to acetate, but they do not have hydrogenases which inhabilite the hydrogen use, so they can produce three molecules of acetate from fructose. Acetic acid is equally a co-metabolite of the organic substrates fermentation (sugars, glycerol, lactic acid, etc.) by diverse groups of microorganisms which produce different acids: propionic bacteria (propionate + acetate);

- *Clostridium* (butyrate + acetate);
- Enterobacteria (acetate + lactate); and
- Hetero-fermentative bacteria (acetate, propionate, butyrate, valerate, etc.).

Anaerobic Respiration

Anaerobic respiration refers to the oxidation of molecules in the absence of oxygen to produce energy, in opposition to Aerobic respiration which does use oxygen. Anaerobic respiration processes require another electron acceptor to replace oxygen. Anaerobic respiration is often used interchangeably with fermentation, especially when the glycolytic pathway is used for energy production in the cell. They are not synonymous terms, however, since certain anaerobic prokaryotes can generate all of their ATP using an electron transport system and ATP synthase. The word and symbol equation for the anaerobic respiration of glucose is:

Glucose $\rightarrow$ Lactic acid + Energy (ATP)

Carbon dioxide (CO_2) is not produced in this type of respiration.

During excercise, this is how we feel cramp or stitches.

$$C_6H_{12}O_6 \rightarrow 2C_3H_6O_3 + 2\ ATP$$

The energy released is about 120kJ per mole Glucose.

Obligate Anaerobes

In some organisms called *obligate (strict) anaerobes* (ex: *C. tetani* (causes tetanus), *C. perfringens* (causes gangrene)), the presence of oxygen is lethal. This is because the presence of oxygen is processed by the organisms into the extremely toxic molecules of single oxygen ($^1O_2$), superoxide ion (O_2^-), hydrogen peroxide (H_2O_2), hydroxyl ion (OH^-), and other toxic molecules

Faculative Anaerobes and Obligate Aerobes

Faculative anaerobes (organisms that can survive in either oxygenated or deoxygenated environments and can switch between cellular respiration or fermentation, respectively) and *obligate (strict) aerobes* (organisms that can survive only with oxygen) have special enzymes (superoxide dimutase and catalase) that can safely handle these products and transform them into harmless water and diatomic oxygen in the following reactions:

1. $2O_2^- + 2H^+ \xrightarrow{\text{Dismutase}} H_2O_2$ (hydrogen peroxide) + O_2

The hydrogen peroxide produced is then transferred to a second reaction...

2. $2H_2O_2 \xrightarrow{\text{Catalase}} 2H_2O + O_2$

The oxidative powers of the superoxide ion have now been neutralised. Only faculative anaerobes and obligate aerobes possess the two enzymes necessary to reduce the superoxide.

In organisms which use glycolysis, the absence of oxygen prevents pyruvate from being metabolised to CO_2 and water via the citric acid cycle and the electron transport chain (which relies on O_2) does not function. Fermentation does not yield more energy than that already obtained from glycolysis (2 ATPs) but serves to regenerate NAD^+ so glycolysis can continue. Various end products can also be created, such as lactate or ethanol.

Fermentation in animals is essential to human life.

In lactic acid fermentation, the following reaction occurs:

1. $C_6H_{12}O_6$ (glucose) + 2 NAD^+ $\longrightarrow$ 2 $C_3H_4O_3$ (pyruvic acid) + 2 NADH

Fermentation in other Organisms

In some plant cells and yeasts, fermentation produces CO_2 and ethanol. The conversion of pyruvate to acetaldehyde generates CO_2 and the conversion of acetaldehyde to ethanol regenerates NAD^+.

Anaerobic Respiration in Prokaryotes

In the field of prokaryotic metabolism, anaerobic respiration has a more specific meaning. In this case, anaerobic respiration is defined as a membrane bound biological process coupling the oxidation of electron donating substrates (e.g. sugars and other organic compounds, but also inorganic molecules like hydrogen, sulphide/sulphur, ammonia, metals or metal ions) to the reduction of suitable *external* electron acceptors other than molecular oxygen.

In contrast, in fermentation the oxidation of molecules is coupled to the reduction of an internally-generated electron acceptor, usually pyruvate. Hence, scientists who study prokaryotic physiology view anaerobic respiration and fermentation as distinct processes, and therefore do not use the terms interchangeably.

In anaerobic respiration, as the electrons from the electron donor are transported down the electron transport chain to the terminal electron acceptor, protons are translocated over the membrane from "inside" to "outside", establishing a concentration gradient across the membrane which temporarily stores the energy released in the chemical reactions. This potential energy is then converted into ATP by the same enzyme used during aerobic respiration, ATP synthase. Possible electron acceptors for anaerobic respiration are nitrate, nitrite, nitrous oxide, oxidised amines and nitro-compounds, fumarate, oxidised metal ions, sulphate, sulphur, sulphoxo-compounds, halogenated organic compounds, selenate, arsenate, bicarbonate or carbon dioxide (in acetogenesis and methanogenesis). All these types of anaerobic respiration are restricted to prokaryotic organisms.

Examples of Anaerobic Respiration:

Glucose + $3NO_3^-$ + $3H_2O \rightarrow$ $6HCO_3^-$ + $3NH_4^+$, $\Delta G^{0'} = -1796$ kJ

Glucose + $3SO_4^{2-}$ + $3H^+ \rightarrow$ $6HCO_3^-$ + $3NH^-$, $\Delta G^{0'} = -453$ kJ

Glucose + 12S + $12H_2O \rightarrow$ $6HCO_3^-$ + $12HS^-$ + $18H^+$, $\Delta G^{0'} = -333$ kJ

All of these terminal electron acceptors are further upstream in the electron transport chain, compared to O_2. Consequently, anaerobic respiration is less effective than aerobic respiration. The ÄG$^{0'}$ of aerobic respiration is -2844 kJ.

Biogas Powerplant

A biogas powerplant is a system where biogas is used to generate electricity. The gas which is produced via anaerobic digestion is used to drive an electricity generator. By-products of this process are steam and hot water. The hot water can be recycled in a combined heat and power cycle to increase the temperature of the digesters to optimal conditions.

Biogas power plants are a combination of anaerobic digestion systems with associated electricity generators such as gas turbines or gas engines. The electricity they produce is classified as renewable or green energy and if sold into the national grid may attract subsidies (such as Renewables Obligation Certificates in the UK).

Feedstock into the biogas power plants must be biodegradable in order to produce methane. Suitable feedstocks include (but are not limited to):

- Biodegradable waste
- Sewage treatment sludge (primary or raw sludge and/ or secondary sludge)
- Slaughterhouse waste
- Food waste
- Farm waste
- Organic component of mixed municipal waste (in mechanical biological treatment)
- Biomass like maize.

There are three stages of anaerobic digestion: hydrolysis, acidogenesis, and methanogenesis. These stages can occur in the same digestion tank or can be controlled independently and optimised according to the requirements of the different bacterial processes.

The more complex and efficient a biogas plant the more expensive it will be for the locality. Biogas plants can be simplified to produce gas for villages in countries where organic wastes are available and funds are limited. Alternatively, in more developed countries pressure in the form of legislation and high energy costs is increasing the amount of projects generating renewable energy from waste.

Biogas plants can be found in countries such as India, China, Philippines, Germany, Austria and Turkey.

Advanced processing systems can recover the organic fraction mixed waste streams. These systems are a subgroup of mechanical biological treatment plants. They sort the recyclable elements of the waste and process the organic fraction into a high surface area low solids soup which are then passed into a biogas power plant (anaerobic digester). Advanced systems like this can be found in Israel.

Further energy can be produced by the combustion of the digestate which may be classified as a biofuel.

Principal Procedure

Biogas Production

- Preparing the biomass
- Mixing
- Digester/fermenter (Heating 40-90°C)

Gas Input

- Raw biogas input 40°C
- Liquid gas Separator '! condensates
- Gas dryer (refrigerate) 4°C condensates

- Gas compressor about 400kPa
- Gas filter (cleaning of dust particles, less than 5 parts per billion by mass of siloxanes)
- Gas heating (minimum about 10°C)

Gas combustion → thermal energy

- Gas turbine exhaust output 300 to 400°C
- Generator → electric energy

Exhaust output → heat exchange

- HRSG heat recovery steam generation
- Heat exchanger for hot water

Siloxane might be present in the biogas and must be removed prior to input in the gas engines, as it erodes moving parts. Hydrogen sulphide may also be produced in the process if there are high levels of sulphur in the biogas. The exhaust gas must be cleaned up, as sulphur dioxide is toxic.

Plant Engines

- Gas motor engines
- Micro–gas turbines
- Fuel cells
- Boiler

Plant Types

Plant type depends on the type of biogas and usage of energy.

- CHP combined heat and power or HRSG,
- CHCP combined heat, cooling and power production,

Plant Sizes

Size	*Power*	*Plant Size*
Small	500W to 5 kW	10 m²
Medium	5 kW to 75 kW	15 to 100 m²
Large	75 kW to 4MW	1 km²

The plant can also be segmented including gas motors and gas turbines.

Mechanical Biological Treatment

Mechanical biological treatment ("MBT"), or "mechanical biological pretreatment" is a category of waste treatment technologies that enables recovery of the resources contained in waste.

The resources, including metals, plastics and glass can then be recycled. It consists of any number of combinations of mechanical sorting and biological treatment processes for mixed waste streams. MBT generally, but not exclusively is used to process unsorted household waste.

MBT is also sometimes termed BMT – biological mechanical treatment – however this simply refers to the order of processing. MBT should not be confused with MHT – *mechanical heat treatment*.

Mechanical Factor

The "mechanical" element is usually an automated mechanical sorting stage. This either removes recyclable elements from a mixed waste stream (such as metals, plastics and glass) or processes them. It typically involves factory style conveyors, industrial magnets, eddy current separators, trommels, shredders and other tailor made systems. The mechanical element has a number of similarities to a materials recovery facility (MRF).

Some systems integrate a wet MRF to recover and wash the recyclable elements of the waste in a form that can be sent for recycling. MBT can alternatively process the waste to produce a high calorific fuel given the term refuse derived fuel (RDF). RDF can be used in cement kilns or power plants and is generally made up from plastics and biodegradable organic waste. Systems which are configured to produce RDF include the Herhof and Ecodeco Processes. It is a common misconception that all MBT processes produce RDF. This is not the case and depends strictly on system configuration and suitable local markets for MBT outputs.

Biological Factor

The "Biological" element refers to either:

- Anaerobic digestion
- Composting

Anaerobic digestion breaks down the biodegradable component of the waste to produce biogas and soil improver. The biogas can be used to generate renewable energy. Some processes through water based processing enable high rates of gas and green energy generation without the production of RDF. This is facilitated by processing the waste in water.

Biological can also refer to a composting stage. Here the organic component is treated with aerobic microorganisms. They break down the waste into carbon dioxide and compost. There is no green energy produced by systems employing only composting treatment for the biodegradable waste.

Some systems, such as UR-3R involve both a partial anaerobic digestion stage along with a secondary composting stage.

By processing the biodegradable waste either by anaerobic digestion or by composting MBT technologies help to reduce the contribution of greenhouse gases to global warming.

Implications for Kerbside Collection

More advanced MBT solutions can diminish the need for home separation and kerbside collection of recyclable elements of waste. This gives the ability of local authorities and councils to reduce the use of waste vehicles on the roads and keep recycling rates high.

This goes against some environmental thinking which pushes for people to recycle in the home so they are more aware and take ownership of the waste they produce.

MBT is gaining increased recognition in countries with changing waste management markets such as the UK.

Friends of Earth Position

Friends of the Earth suggests that the best environmental route for residual waste is to firstly remove the remaining recyclable elements from the waste stream (such as metals, plastics and some paper). The small amount of waste remaining should be composted or anaerobically digested and unless sufficiently clean to be used as a compost should be disposed of in landfill. MBT plants that fit in with these statements could therefore play an increasing role in the environmental management of mixed streams.

Process Examples

ArrowBio

Water-based MBT; ArrowBio, is a variant technology that comes under the general category of *mechanical biological treatment*, however is distinct in its application of water. The system combines mechanical sorting in a water-based materials recovery facility with water-based UASB anaerobic digestion.

The system washes and extracts the recyclable elements of the unsorted bin bags recovering resources such as metals, plastics and glass. This process does not produce RDF and does not require its outputs to be burnt. The technology produces green energy from biogas and clean soil improver. A "hydrocrushing" stage results in significantly more of the biodegradable material being converted into biogas.

A commercially operating reference facility is based at the Hiriya waste park. A second system is under construction at the Jack's Gulley landfill and will be operated by WSN Environmental Solutions.

Process Inputs:

- Mixed municipal solid waste (or similar biodegradable waste stream).

Process Outputs:

- Cleaned recyclables: Ferrous metal, non-ferrous metals, HDPE, PET, LDPE.

- Biogas (converted into renewable energy).
- Soil improver (digestates from anaerobic digestion).
- A residue of non-recoverable elements which is sent to landfill (c20 per cent of input).
- Cleaned water recirculated in system and exported.

Biochemical Oxygen Demand (BOD)

Biological oxygen demand is a chemical procedure for determining how fast biological organisms use up oxygen in a body of water. It is used in water quality management and assessment, Ecology and environmental science. BOD is not an accurate quantitative test, although it could be considered as an indication of the quality of a water source.

Typical BOD Values

Most pristine rivers will have a 5 day BOD of less than 1 mg/l. Moderately polluted rivers may have BODs in the range 2 mg/l to 8 mg/l. Efficiently treated municipal sewage treated by a three stage process would have a BOD value of about 20 mg/l. Untreated sewage is variable but might average out at around 600 mg/l in Europe and as low as 200 mg/l in the USA or where there is severe groundwater or surface water infiltration. (The generally lower values in the USA derive from the much greater water use per capita than in other parts of the world.) Slurry from dairy farms is around 8000 mg/l and silage liquor around 60000 mg/l.

BOD Level (in ppm)	*Water Quality*
1 - 2	Very Good-not much organic waste present
3 - 5	Moderately clean
6 - 9	Somewhat polluted
10+	Gigimax polluted

The BOD Test

BOD measures the rate of oxygen uptake by microorganisms in a sample of water at a fixed temperature (20°C) and over a given period of time(usually five days) in the dark. To ensure that all other conditions are equal, a very small amount of microorganism seed is added to each sample being tested. This seed is typically generated by diluting activated sludge with de-ionised water. The test generally takes place over an elapsed period of five days but other BOD tests are also used.

Method

The BOD test is carried out by diluting the sample with de-ionised water saturated with oxygen, inoculating it with a fixed aliquot of seed, measuring the dissolved oxygen and sealing the sample (to prevent further oxygen dissolving in). The sample is kept at 20°C in the dark (to prevent photosynthesis and thereby the addition of oxygen) for five days and the dissolved oxygen DO is measured again. The difference between the final DO and initial DO is the BOD. The apparent BOD for the control is subtracted from the control result to provide the corrected value.

The loss of dissolved oxygen in the sample, once corrections have been made for the degree of dilution, is called the BOD5. In the UK allylthiourea is also added at the start of the test to prevent oxidation of ammonia. Results from such tests are represented as BOT5(ATU) and referred to as Carbonaceous BOD (CBOD) in the US. Less frequently used is the Ultimate BOD (UBOD) test, in which DO is repeatedly measured by DO metre in the same specialised bottles until it has reached equilibrium.

BOD is similar in function to chemical oxygen demand (COD), in that both measure the amount of organic compounds in water. However, COD is less specific since it measures everything that can be chemically oxidised rather than just levels of biology active organic matter.

BOD is used as a gauge of the effectiveness of wastewater treatment plants. Various commercial devices are available for its determination.

BOD can be calculated by: Undiluted; Initial D.O - Final

D.O = BOD Diluted; ((Initial D.O - Final D.O)- BOD of Seed) x Dilution Factor

History of the Use of BOD

The *Royal Commission on River Pollution* which was established in 1865 and the formation of the *Royal Commission on Sewage Disposal* in 1898 led to the selection in 1908 of BOD5 as the definitive test for organic pollution of rivers. Five days was chosen as an appropriate test period because this is supposedly the longest time that river water takes to travel from source to estuary in the UK. In 1912, the commission also set a standard of 20 ppm BOD5 as the maximum concentration of permitted in sewage works discharging to rivers provided that there was at least an 8:1 dilution available at dry weather flow. This was contained in the famous 20:30 (BOD:Suspended Solids) + full nitrification standard which was used as a yardstick in the UK up to the 1970s for sewage works effluent quality.

Bioconversion of Biomass to Mixed Alcohol Fuels

The bioconversion of biomass to mixed alcohol fuels can be accomplished using the MixAlco process. Through bioconversion of biomass to a mixed alcohol fuel, more energy from the biomass will end up as liquid fuels than in converting biomass to ethanol by yeast fermentation.

The process involves a biological/chemical method for converting any biodegradable material (e.g., urban wastes, such as municipal solid waste, biodegradable waste, and sewage sludge, agricultural residues such as corn stover, sugarcane bagasse, cotton gin trash, manure) into useful chemicals, such as carboxylic acids (e.g., acetic, propionic, butyric acid), ketones (e.g., acetone, methyl ethyl ketone, diethyl ketone) and biofuels, such as a mixture of primary alcohols (e.g., ethanol, propanol, butanol) and/or a mixture of secondary alcohols (e.g., isopropanol, 2-butanol, 3-pentanol). Because of the many products that can be economically produced, this process is a true biorefinery.

The process uses a mixed culture of naturally occurring microorganisms found in natural habitats such as the rumen of cattle, termite guts, and marine and terrestrial swamps to anaerobically digest biomass into a mixture of carboxylic acids produced during the acidogenic and acetogenic stages of anaerobic digestion, however with the inhibition of the methanogenic final stage. The more popular methods for production of ethanol and cellulosic ethanol use enzymes that must be isolated first to be added to the biomass and thus convert the starch or cellulose into simple sugars, followed then by yeast fermentation into ethanol. This process does not need the addition of such enzymes as these microorganisms make their own.

As the microoganisms anaerobically digest the biomass and convert it into a mixture of carboxylic acids, the pH must be controlled. This is done by the addition of a buffering agent (e.g., ammonium bicarbonate, calcium carbonate), thus yielding a mixture of carboxylate salts. Methanogenesis, which, as mentioned, is the natural final stage of anaerobic digestion, is inhibited by the presence of the ammonium ions or by the addition of an inhibitor (e.g., iodoform). The resulting fermentation broth contains the produced carboxylate salts that must be dewatered. This is achieved efficiently by vapour-compression evaporation. Further refining of the dewatered fermentation broth may then take place depending on the final product desired.

The condensed distilled water from the vapour-compression evaporation system is recycled back to the fermentation. On the other hand, if sewage or other waste water with high BOD in need of treatment is used as the water for the fermentation, the condensed distilled water from the evaporation can be recycled back to the city or to the original source of the high-BOD waste water. Thus, this process can also serve as a water treatment facility.

Because the system uses a mixed culture of microorganisms, besides not needing any enzyme addition, the fermentation requires no sterility or aseptic conditions, making this front step in the process more economical than in more popular methods for the production of cellulosic ethanol. These savings in the front end of the process, where volumes are large, allows flexibility for further chemical transformations after dewatering, where volumes are small.

Carboxylic Acids

Carboxylic acids can be regenerated from the carboxylate salts using a process known as "acid springing". This process makes use of a high-molecular-weight tertiary amine (e.g., trioctylamine), which is switched with the cation (e.g., ammonium or calcium). The resulting amine carboxylate can then be thermally decomposed into the amine itself, which is recycled, and the corresponding carboxylic acid. In this way, theoretically, no chemicals are consumed or wastes produced during this step.

Ketones

There are two methods for making ketones. The first one consists on thermally converting calcium carboxylate salts into the corresponding ketones. This was a common method for making acetone from calcium acetate during World War I. The other method for making ketones consists on converting the vaporised carboxylic acids on a catalytic bed of zirconium oxide.

Alcohols

Primary Alcohols

The undigested residue from the fermentation may be used in gasification to make hydrogen (H_2). This H_2 can then be used to hydrogenolyse the esters over a catalyst (e.g., copper chromite), which are produced by esterifying either the ammonium carboxylate salts (e.g., ammonium acetate, propionate, butyrate) or the carboxylic acids (e.g., acetic, propionic, butyric acid) with a high-molecular-weight alcohol (e.g., hexanol, heptanol). From the hydrogenolysis, the final products are the high-molecular-weight alcohol, which is recycled back to the esterification, and the corresponding primary alcohols (e.g., ethanol, propanol, butanol).

Secondary Alcohols

The secondary alcohols (e.g., isopropanol, 2-butanol, 3-pentanol) are obtained by hydrogenating over a catalyst (e.g., Raney nickel) the corresponding ketones (e.g., acetone, methyl ethyl ketone, diethyl ketone).

Acetic Acid vs. Ethanol

Cellulosic-ethanol-manufacturing plants are bound to be net exporters of electricity because a large portion of the lignocellulosic biomass, namely lignin, remains undigested and it must be burned, thus producing electricity for the plant and excess electricity for the grid. As the market grows and this technology becomes more widespread, coupling the liquid fuel and the electricity markets will become more and more difficult.

Acetic acid, unlike ethanol, is biologically produced from simple sugars without the production of carbon dioxide:

$$C_6H_{12}O_6 \rightarrow 2\ CH_3CH_2OH + 2\ CO_2$$

(Biological production of ethanol)

$$C_6H_{12}O_6 \rightarrow 3\ CH_3COOH$$

(Biological production of acetic acid)

Because of this, on a mass basis, the yields will be higher than in ethanol fermentation. If then, the undigested residue (mostly lignin) is used to produce hydrogen by gasification, it is ensured that more energy from the biomass will end up as liquid fuels rather than excess heat/electricity.

$$3\ CH_3COOH + 6\ H_2 \rightarrow 3\ CH_3CH_2OH + 3\ H_2O$$

(Hydrogenation of acetic acid)

$$C_6H_{12}O_6 \text{ (from cellulose)} + 6\ H_2 \text{ (from lignin)} \rightarrow 3\ CH_3CH_2OH + 3\ H_2O$$

(Overall reaction)

Stage of Development

The system has been in development since 1991, moving from the laboratory scale (10 g/day) to the pilot scale (200 lb/day) in 1998. A 100 ton/day demonstration plant is expected in 2009.

Biosolids

Biosolids is a public relations term used by the sewage sludge industry to greenwash toxic sludge. Sludge, or "biosolids," are the by-product of the treatment of domestic wastewater in a wastewater treatment plant. To create biosolids, these residuals are further treated to reduce pathogens and vector attraction by any of a number of approved methods. Nevertheless, toxic chemicals, such as PCBs, dioxin, and brominated flame retardants, remain in the "treated" sludge, as there is no technology available to remove these and tens of thousands of other chemicals from sewage sludge, the by-product of wastewater treatment. The more treated the wastewater, the more toxic the sludge. Depending on their level of treatment and resultant pollutant content, biosolids can be used in regulated applications ranging from soil conditioning to fertilizer for food or non-food agriculture to distribution for unlimited use. This is a result of EPA stonewalling and disregard for science in its analysis of the safety of land application of sewage sludge.

The term biosolids was formally recognised in 1991 by the Water Environment Federation (WEF). WEF, founded in 1928, is a not-for-profit technical and educational organisation with members from varied disciplines (e.g. wastewater treatment operators and engineers) who work for the preservation and enhancement of the global water environment. WEF was formerly known as the "Federation of Sewage Works Associations." Biosolids is the term created in 1991 by the *Name Change Task Force* at WEF to differentiate raw, untreated sewage sludge from treated and tested sewage sludge that can be beneficially utilised as soil amendment and fertilizer. The term "biosolids" also helps make the land application of processed sewage sludge more acceptable to the public. The proposal to create a "Name Change Task Force" originated with Peter Machno, manager of Seattle's sludge programme, after protesters mobilised against his plan to spread sludge on local tree farms. "If I knocked on your door and said I've got this beneficial product called sludge, what are you going to say?" he asked. At Machno's suggestion, the Federation newsletter published a request for alternative names. Members sent in over 250 suggestions, including "all growth," "purenutri," "biolife," "bioslurp," "black gold," "geoslime," "sca-doo," "the end product," "humanure," "hu-doo," "organic residuals," "bioresidue," "urban biomass," "powergro," "organite," "recyclite," "nutri-cake" and "ROSE," short for "recycling of solids environmentally." In June of 1991, the Name Change Task Force finally settled on "biosolids," which it defined as the "nutrient-rich, organic by-product of the nation's wastewater treatment process."

The new name attracted sarcastic comment from the Doublespeak Quarterly Review, edited by Rutgers University professor William Lutz. "Does it still stink?" Lutz asked. He predicted that the new name "probably won't move into general usage. It's obviously coming from an engineering mentality. It does have one great virtue, though. You think of `biosolids' and your mind goes blank."

According to Machno, the name change was not intended to "cover something up or hide something from the public.... We're trying to come up with a term... that can communicate to the public the value of this product that we spend an awful lot of money on turning into a product that we use in a beneficial way."

Processes

During waste water treatment, bacteria and other microorganisms break down components in wastewater into simpler and more stable forms of organic matter. Non-organic matter also settles into sludge. For instance, small amounts (parts per million) of heavy metals and other potentially toxic materials, including flame retardants (PBDEs) and persistent organic pollutants, are commonly found in sewage sludge in parts per million levels. What does not settle into sludge leaves the treatment facility as a treated wastewater effluent. Biosolids in their liquid form look like muddy water and contain 1-10 per cent solids. Biosolids may be dewatered in a second step of the treatment process, which turns it into a "cake" with the texture of a wet sponge. In this stage, it contains 11-40 per cent solids.

Biosolids in USA

According to US EPA, biosolids that meet treatment and pollutant content criteria "can be safely recycled and applied as fertilizer to sustainably improve and maintain productive soils and stimulate plant growth." After the 1991 Congressional ban on ocean dumping, the US EPA promulgated regulations—40 CFR Part 503—that continued to allow for the use of biosolids on land as fertilizers and soil amendments (prior regulations, the Part 257, had allowed such uses). This is a highly controversial policy decision that has been contested since its inception. EPA promoted biosolids recycling throughout the 1990s. The EPA's Part 503 regulations were developed with input from university, EPA, and USDA researchers from around the country and involved an extensive review of the scientific literature and the largest risk assessment the agency had conducted to that time. Still, many academics, government researchers, and citizens contest EPA's conclusions about the safety of land applying sewage sludge. The Part 503 regulations became effective in 1993.

In the United States municipal wastewater treatment plants annually produced about 7.7 million dry tons of biosolids in 1997, and about 6.8 million dry tons in 1998 according to sources relying on USEPA estimates. According to the NRC, about 5.6 million dry tons was the normal US annual biosolids production rate as of 2002.

In the United States, as of 2002, about 60 per cent of all

biosolids are applied to land as soil amendment and fertilizer for growing crops. Biosolids that meet the Class B pathogen treatment and pollutant criteria, in accordance with the US EPA "Standards for the use or disposal of sewage sludge," can be land applied with formal site restrictions and strict record keeping. Biosolids with lower pollutant content have fewer restrictions. Biosolids that meet Class A pathogen reduction requirements or equivalent treatment by a Process to Further Reduce Pathogens (PFRP) have the least restrictions on use. PFRPs include composting, heat drying, heat treatment, thermophilic aerobic digestion, beta or gamma ray irradiation and pasteurisation. Processes to reduce pathogens have no effect on heavy metals and may or may not have effects on the levels of other trace pollutants in biosolids.

EPA policy on sewage sludge recycling is controversial. Often thought to consist of only "human waste," treated sewage sludge or "biosolids" contains any contaminants from sewage that are not broken down in the treatment process or which do not remain with the water effluent leaving the treatment plant. The most commonly detected trace contaminants of concern are heavy metals (arsenic, cadmium, copper, etc.) and toxic chemicals (e.g. widely used plasticisers, PDBEs, etc.). Pathogens are also a significant health issue.

Over the years there have been thousands of reported incidents of harm caused by biosolids use on land. Symptoms have characteristically included: asthma, weight loss, fatigue, eye irritations, flu-like symptoms, gastrointestinal complications, headaches, immunodeficiency problems, lesions, nausea, nosebleeds, rashes, respiratory complications, abscesses, reproductive complications, cysts, and tumours. Currently, there is no formal tracking system for complaints related to sewage sludge. EPA has never conducted an epidemiological study, though thousands of people have complained of health problems as a result of living near land applied sites. Some deaths have been reported.

The EPA categorically denies that land applied sewage

sludge is harmful. But the claim of harm to animals and land was recently litigated to judgment by at least one court in the United States. On June 24, 2003, after reviewing detailed evidence, a Richmond County, Georgia jury found that sewage sludge from the city was responsible for killing cows and damaging farmland at the Boyce family farm.

The National Research Council published "Biosolids Applied to Land: Advancing Standards and Practices" in July 2002. They concluded that there is no documented scientific evidence that biosolids regulations have failed to protect public health, but there is persistent uncertainty on possible adverse health effects. The NRC noted that further research is needed and made about 60 recommendations for addressing public health concerns, scientific uncertainties, and data gaps in the science underlying the sewage sludge standards. EPA responded with commitment to conduct research addressing the most important, but not all, of the NRC recommendations.

The EPA Office of the Inspector General (OIG) completed two assessments in 2000 and 2002 of the EPA sewage sludge programme. The 2000 report declared that the EPA did not have an "effective programme for ensuring compliance with the land application requirements of the Sludge Rule." This report further documented that EPA performed "virtually no inspections" of land application sites and few inspections of treatment plants or land appliers." The report also documented the lack of resources committed to sludge and the low priority placed on the sludge programme by the EPA.

The follow-up report in 2002 documented that "EPA cannot assure the public that current land application practices are protective of human health and the environment." The report also documented that there had been an almost 50 per cent reduction in EPA enforcement resources since the earlier assessment. The report documented EPA's failure to create formal processes for tracking and responding to human health complaints related to land application of sludge.

Biosolids in the European Union

In Europe, the recycling of biosolids (treated sewage sludge) to agriculture is practised widely. It is supported by the European Commission and many European governments and, in most circumstances, this beneficial reuse is regarded as the best environmental option. Over 40 per cent of all sewage sludge produced in the EU is recycled to agricultural land (based on data from the European Commission from 1999-2000) although the actual rate varies considerably from country to country. In countries such as France, Spain, the UK, Denmark and Luxemburg over 50 per cent of production is recycled to land.

The process is well regulated, though highly controversial, principally under the 1986 EU Sludge Directive (86/27/EEC). The main objective of the Directive is the control of heavy metals (PTEs), thought to be the main contaminants of concern when biosolids are being applied to land. The regulatory controls introduced address the potential impact of biosolids addition to soil from both the immediate application and also any possible cumulative effects.

In the UK, the EU Sludge Directive is implemented through the Sludge (Use in Agriculture) Regulations 1989. These Regulations are supported by a detailed Code of Practice that describes all aspects of biosolids recycling to land. The regulations set permissible limits for soil concentrations and rates of annual additions of PTEs. The allowable limits for Zn, Cu and Ni in soils vary with the pH of the soil.

Since 1998 the UK water industry has also complied with the additional requirements of the Safe Sludge Matrix. This is a voluntary agreement made between the UK water and sewerage operators, the British Retail Consortium (BRC), representing the major retailers, and a range of other stakeholders. The matrix introduced strict controls on the microbiological quality of sludge and specified procedures to be adopted for its application to agricultural land used to grow food crops. The use of untreated sludge on agricultural land for food production was phased out in 1999 and the use of untreated sludge on agricultural land used to grow non-food crops was phased out in 2005. The provisions of the Matrix go beyond the requirements of the Sludge (Use in) Agriculture Regulations as they currently stand.

Water Companies in the UK have also adopted HACCP (Hazard Analysis and Critical Control Point) procedures for sludge stream management. HACCP procedures apply risk assessment and process control to manage and reduce risk, ensuring that the pathogen reduction requirements specified by the Safe Sludge Matrix are met, formalising record keeping and maintaining quality control.

Further legislation is on the horizon in the form of a revision to the 1986 EU Sludge Directive. The Commission's intention is to propose guidelines on sustainable practices for the application of treated sludge (biosolids) onto land. In this context, it will also be assessed whether the scope of the revised directive should be broadened to other non-hazardous sludges and to applications other than in agriculture. The aim of the revision to the Directive will be, on the one hand, to encourage the sustainable use of properly treated sludge and, on the other, to strengthen the controls applied in order to guarantee that both professional users and the public in general have increased confidence in the practice of recycling biosolids to land.

Chemical Oxygen Demand

In environmental chemistry, the chemical oxygen demand (COD) test is commonly used to indirectly measure the amount of organic compounds in water. Most applications of COD determine the amount of organic pollutants found in surface water (e.g. lakes and rivers), making COD a useful measure of water quality. It is expressed in milligrams per litre (mg/L), which indicates the mass of oxygen consumed per litre of solution. Older references may express the units as parts per million (ppm).

The basis for the COD test is that nearly all organic compounds can be fully oxidised to carbon dioxide with a strong oxidising agent under acidic conditions. The amount of oxygen required to oxidise an organic compound to carbon dioxide, ammonia, and water is given by:

$$C_nH_aO_bN_c\left(n+\frac{a}{4}-\frac{3}{4}c\right)O_2 \rightarrow nCO_2+\left(\frac{a}{2}-\frac{3}{2}c\right)H_2O+cNH_3$$

This expression does not include the oxygen demand caused by the oxidation of ammonia into nitrate. The process of ammonia being converted into nitrate is referred to as nitrification. The following is the correct equation for the oxidation of ammonia into nitrate.

$$NH_3+2O_2 \rightarrow NO_3^- + H_3O^+$$

The second equation should be applied after the first one to include oxidation due to nitrification if the oxygen demand from nitrification must be known. Dichromate does not oxidise ammonia into nitrate, so this nitrification can be safely ignored in the standard chemical oxygen demand test.

The International Organisation for Standardisation describes a standard method for measuring chemical oxygen demand in ISO 6060.

History

For many years, the strong oxidising agent potassium permanganate ($KMnO_4$) was used for measuring chemical oxygen demand. Measurements were called *oxygen consumed* from permanganate, rather than the *oxygen demand* of organic substances. Potassium permanganate's effectiveness at oxidising organic compounds varied widely, and in many cases biochemical oxygen demand (BOD) measurements were often much greater than results from COD measurements. This indicated that potassium permanganate was not able to effectively oxidise all organic compounds in water, rendering it a relatively poor oxidising agent for determining COD.

Since then, other oxidising agents such as ceric sulphate, potassium iodate, and potassium dichromate have been used to determine COD. Of these, potassium dichromate ($K_2Cr_2O_7$) has been shown to be the most effective: it is relatively cheap, easy to purify, and is able to nearly completely oxidise almost all organic compounds.

Using Potassium Dichromate

Potassium dichromate is a strong oxidising agent under acidic conditions. (Acidity is usually achieved by the addition of sulphuric acid.) The reaction of potassium dichromate with organic compounds is given by:

$$C_nH_oO_bN_c + C_{r_2}O_7^{2-} + (8d+c)H+ \rightarrow nCO_2 + \frac{a+8d-3c}{2}$$

$$H_2O + cNH_2^+ + 2dCr^3 +$$

where $d = 2n/3 + a/6 - b/3 - c/2$. Most commonly, a 0.25 N solution of potassium dichromate is used for COD determination, although for samples with COD below 50 mg/L, a lower concentration of potassium dichromate is preferred.

In the process of oxidising the organic substances found in the water sample, potassium dichromate is reduced (since in all redox reactions, one reagent is oxidised and the other is reduced), forming Cr^{3+}. The amount of Cr^{3+} is determined after oxidisation is complete, and is used as an indirect measure of the organic contents of the water sample.

Blanks

Because COD measures the oxygen demand of organic compounds in a sample of water, it is important that no outside organic material be accidentally added to the sample to be measured. To control for this, a so-called blank sample is required in the determination of COD (and BOD, for that matter). A blank sample is created by adding all reagents (e.g. acid and oxidising agent) to a volume of distilled water. COD is measured for both the water and blank samples, and the two are compared. The oxygen demand in the blank sample is subtracted from the COD for the original sample to ensure a true measurement of organic matter.

Measurement of Excess

For all organic matter to be completely oxidised, an excess amount of potassium dichromate (or any oxidising agent) must be present. Once oxidation is complete, the amount of excess potassium dichromate must be measured to ensure that the amount of Cr^{3+} can be determined with accuracy. To do so, the excess potassium dichromate is titrated with ferrous ammonium sulphate (FAS) until all of the excess oxidising agent has been reduced to Cr^{3+}. Typically, the oxidation-reduction indicator Ferroin is added during this titration step as well. Once all the excess dichromate has been reduced, the Ferroin indicator changes from blue-green to reddish-brown. The amount of ferrous ammonium sulphate added is equivalent to the amount of excess potassium dichromate added to the original sample.

Calculations

The following formula is used to calculate COD:

$$COD = \frac{8000(b-s)n}{\text{sample volume}}$$

where *b* is the volume of FAS used in the blank sample, *s* is the volume of FAS in the original sample, and *n* is the normality of FAS. If millilitres are used consistently for volume measurements, the result of the COD calculation is given in mg/L.

Inorganic Interference

Some samples of water contain high levels of oxidisable inorganic materials which may interfere with the determination of COD. Because of its high concentration in most wastewater, chloride is often the most serious source of interference. Its reaction with potassium dichromate follows the equation

$$6Cl^- + C_{r_2}O_7^{2-} + 14H^+ \rightarrow 3Cl_2 + 2C_{r^{3+}} + 7H_2O$$

Prior to the addition of other reagents, mercuric sulphate can be added to the sample to eliminate chloride interference.

The following table lists a number of other inorganic substances that may cause interference. The table also lists chemicals that may be used to eliminate such interference, and the compounds formed when the inorganic molecule is eliminated.

Inorganic molecule	*Eliminated by*	*Elimination forms*
Chloride	Mercuric sulphate	Mercuric chloride complex
Nitrite	Sulphamic acid	N_2 gas
Ferrous iron	–	–
Sulphides	–	–

Government Regulation

Many governments impose strict regulations regarding the maximum chemical oxygen demand allowed in wastewater before they can be returned to the environment. For example, in Switzerland, a maximum oxygen demand between 200 and 1000 mg/L must be reached before wastewater or industrial water can be returned to the environment.

Composting

Composting is the way of making compost through aerobic decomposition of biodegradable organic matter. The decomposition is performed primarily by although larger creatures such as ants, nematodes, and oligochaete worms also contribute. This decomposition occurs naturally in all but the most hostile environments, such as within landfills, extremely arid deserts or cold weather such as boreal winters or polar regions, which prevent the microbes and other decomposers from thriving.

Composting can be divided into the two areas of home composting and industrial composting. Both scales of composting use the same biological processes, however techniques and different factors must be taken into account.

Composting is the controlled decomposition of organic matter. Rather than allowing nature to take its slow course, a composter provides an optimal environment in which decomposers can thrive. To encourage the most active microbes, a compost pile needs the correct mix of the following ingredients:

- Carbon
- Nitrogen
- Oxygen (in the case of aerobic composting)
- Water

Decomposition happens even in the absence of some of these ingredients, but not as quickly or as pleasantly. (For example, vegetables in a plastic bag will decompose, but the absence of air encourages the growth of anaerobic microbes, which produce disagreeable odours. Degradation under anaerobic conditions is called anaerobic digestion.)

The Goal of a Composting System

The goal in a composting system is to provide a healthy environment and nutrition for the rapid decomposers, the bacteria. The most rapid composting occurs with the ideal carbon to nitrogen ratio of between 25 and 30 to 1 by dry chemical weight. In other words, the ingredients placed in the pile should contain 25 to 30 times as much carbon as nitrogen. For example, grass clippings average about 19 to 1 and dry autumn leaves average about 55 to 1. Mixing equal parts by volume approximates the ideal range. Commercial-grade composting operations pay strict attention to this ratio. For backyard composters, however, the charts of carbon and nitrogen ratios in various ingredients and the calculations required to get the ideal mixture can be intimidating, so many rules of thumb exist to guide composters in approximating this mixture.

Materials for Composting

Given enough time, all biodegradable material will compost. However, not all compost feedstocks are appropriate for backyard composting. Most backyard systems will not reach high enough temperatures to kill pathogens and deter vermin, so pet droppings, non-vegetarian animal manure, meat scraps, and dairy products are best left to operators of high-rate, thermophilic composting systems.

Certain substances should not be composted by the average homeowner, as they require more sophisticated systems.

These substances include non-vegetarian animal manures and bedding, by-products of food production and processing, restaurant grease and cooking oils, and residuals from the treatment of wastewater and drinking water. Composting will also break down petroleum hydrocarbons and some toxic compounds for recycling and beneficial reuse. The use of composting for such purposes is most commonly referred to as a form of bioremediation.

High-carbon sources provide the cellulose needed by the composting bacteria for conversion to sugars and heat, while high-nitrogen sources provide the most concentrated protein, which allow the compost bacteria to thrive.

Some Ingredients with Higher Carbon Content:

- Dry, straw-type material, such as cereal straws
- Autumn leaves
- Sawdust and wood chips
- Some paper and cardboard (such as corrugated cardboard or newsprint with soy-based inks).

Some Ingredients with Higher Nitrogen Content:

- Green plant material (fresh or wilted) such as crop residues, hay, grass clippings, weeds
- Animal manures (choose vegetarian horse manure, cow manure, llama manure, etc.)
- Fruit and vegetable trimmings
- Seaweeds
- Coffee grounds.

Poultry manure provides lots of nitrogen but little carbon. Horse manure provides both. Sheep and cattle manure don't drive the compost heap to as high a temperature as poultry or horse manure, so the heap takes longer to produce the finished product.

Mixing the materials as they are added increases the rate of decomposition, but it can be easier to place the materials in alternating layers, approximately 15 cm (6 in) thick, to help estimate the quantities. Keeping carbon and nitrogen sources separated in the pile can slow down the process, but decomposition will occur in any event.

Greasy food waste and wastes from meat, dairy products, and eggs should not be used in household compost because they tend to attract unwanted vermin and they do not appropriately decompose in the time required. However, eggshells are a good source of nutrients for the compost pile and the soil although they typically take more than one year to decompose. If recycling of meat and dairy products is desirable, Bokashi is a suitable alternative, which uses fermentation. However, even in Bokashi, liquids like milk and oil should not be used. Manure from non-vegetarian animals should never be used, and neither should human or pet waste.

Composting Techniques

There are a number of different techniques for composting, all employing the two primary methods of aerobic composting:

- *Active* (or *hot*) composting allows aerobic bacteria to thrive, kills most pathogens and seeds, and rapidly produces usable compost. Aerobic bacteria produce less odour and fewer destructive greenhouse gases than their anaerobic counterparts. In addition, they are usually faster at breaking down material and the faster material is broken down, the faster compost is created for your garden.

Pasteurisation in a hot compost (such as the Compost Oven) will occur in any garden compost bin if the temperature reaches above 55°C (131°F) for three or more days. To achieve it, you need to keep your garden compost bin warm, insulated and damp since this encourages the cultivation of actinomycetes, a vital bacteria in the pasteurisation process.

The pasteurised soil naturally created through heat in the garden compost is valuable for the composting home gardener, since the pasteurisation process is otherwise both expensive and complicated, and adding chemicals to produce the pasteurisation effect makes the compost less healthy.

- *Passive* (or *cold*) composting allows nature to take its course in a more leisurely manner, while leaving many pathogens and seeds dormant in the pile.

Cold composting is the type of composting done in most domestic garden compost bins in which temperatures never reach above 30°C (86°F). Cold composting is characterised by individuals putting their kitchen scraps in the garden compost bin and leaving them untended. This "scrap bin," having a very high moisture content and without aeration, is likely to turn anaerobic and generate foul odours, including significant adverse greenhouse gas emissions.

When composting this way, a gardener can improve the process by adding some wood chips or small pieces of bark, leaves, twigs or a combination of these materials distributed throughout the mixture. This material helps to improve drainage and airflow.

Such composting systems may be either enclosed (home container composting, industrial in-vessel composting) or in exposed piles (industrial windrow composting).

Home Composting

Home composters use a range of techniques, varying from extremely passive composting (throw everything in a pile in a corner and leave it alone for a year or two) to extremely active (monitoring the temperature, turning the pile regularly, and adjusting the ingredients over time). Some composters use mineral powders to absorb smells, although a well-maintained pile seldom has bad odours.

Microbes and Heating the Pile

An effective compost pile is kept about as damp as a well wrung-out sponge. This provides the moisture that all life needs to survive. Bacteria and other microorganisms fall into a variety of groups in terms of what their ideal temperature is and how much heat they generate as they do their work. Mesophilic bacteria enjoy midrange temperatures, from about 20 to 40°C (70 to 110°F). As they decompose the organic matter they generate heat, and the inner part of a compost pile heats up the most.

The heap should be about 1 m (3 ft) wide, 1 m (3 ft) tall, and as long as is practicable. This provides a suitable insulating mass to allow a good heat build-up as the material decays. The ideal temperature is around 60°C (140°F), which kills most pathogens and weed seeds while providing a suitable environment for thermophilic (heat-loving) bacteria, which are the fastest acting decomposers. The centre of the heap can get too warm, possibly hot enough to burn a bare hand. If this fails to happen, common reasons include the following:

- The heap is too wet, thus excluding the oxygen required by the compost bacteria.
- The heap is too dry, so that the bacteria do not have the moisture needed to survive and reproduce.
- There is insufficient protein (nitrogen-rich material).

The solution is to add material, if necessary, and/or to turn the pile to aerate it.

Depending on how quickly the compost is required, the heap can be turned one or more times to bring the outer layers to the inside of the heap and vice versa, as well as to aerate the mixture. Adding water at this time helps keep the pile damp. One guideline is to turn the pile when the high temperature has begun to drop, indicating that the food source for the fastest-acting bacteria (in the centre of the pile) has been largely consumed. When turning the pile does not result in a temperature rise, there is no further advantage in turning the pile. When all the material has turned into dark brown or nearly black crumbly matter, it is ready to use.

Worm Composting

Recycling the organic waste of a household into compost allows us to return badly needed organic matter to the soil. In this way, we participate in nature's cycle, and cut down on garbage going into burgeoning landfills. Worm composting or vermicomposting is a method for recycling food waste into a rich, dark, earth-smelling soil conditioner. The great advantage of worm composting is that this can be done indoors and outdoors, thus allowing year round composting. It also provides apartment dwellers with a means of composting. The worm then excretes a soil-nutrient material called worm castings.

This is why wise farmers have historically wanted to have healthy worm populations living in their fields. Worms are at the bottom level of the food chain, and thus are critical to healthy soil. In a nutshell, worm compost is made in a container filled with moistened bedding and redworms. Add your food waste for a period of time, and the worms and microorganisms will eventually convert the entire contents into rich compost. Some good gardeners have developed a radical composting product, made through a brewing process which runs distilled water through Red Wiggler worm castings.

The nutritious elements and microorganisms of the castings are captured in a concentrated liquid form, named worm tea. By using worm tea on your plants and gardens, you put healthy microorganisms back into the soil where they thrive and multiply, creating a much healthier growing environment for your plants.

Industrial Composting

Industrial composting systems are being increasingly installed as an alternative form of waste management to landfill along with other advanced waste processing systems. The industrial composting or anaerobic digestion can be combined with mechanical sorting of mixed waste streams and is given the term mechanical biological treatment. Industrial composting helps prevent global warming by treatment of biodegradable waste before it enters landfill. If this waste is landfilled it breaks down anaerobically producing landfill gas that contains methane, a potent greenhouse gas.

Most commercial and industrial composting operations use active composting techniques. This ensures a higher quality product and produces results in the shortest time. The greatest control, and therefore the highest quality, is generally achieved by composting inside an enclosed vessel which is monitored and adjusted continuously for optimal temperature, air flow, moisture, and other parameters.

Large-scale composting systems are used by a few urban centres around the world. Co-composting is a technique which combines solid waste with de-watered biosolids. The world's largest co-composter is in Edmonton, Alberta, Canada, which turns 220,000 tons of residential solid waste and 22,500 dry tons of biosolids per year into 80,000 tons of compost using a facility 38,690 square metres in size (equivalent to 8 football fields). The aeration building alone is the largest stainless steel building in North America (the size of 14 NHL rinks).

Other Ingredients

Some users like to put special materials and activators into their compost. Adding commercially available Effective Microorganisms helps to keep the balance between "good" and "bad" bacteria. A light dusting of agricultural lime (not on the animal manure layers) can curb excessive acidity that can slow down the fermentation. Seaweed meal can provide a ready source of trace elements. Finely pulverised rock (rock flour or rock dust) can also provide needed minerals, as opposed to clay (which is trace mineral-poor and/or leached rock dust).

Animal manure should only be collected from vegetarian animals, such as horses, cows, sheep, llamas, etc. Pet waste, human waste, and non-vegetarian animal waste should not be used in the average compost heap.

Human waste can be collected by composting toilets (in this case, human faeces). However, such compost is usually not used as a fertilizer for plants that are directly edible (e.g., salad crops). Most composting toilets do not allow for the thermophilic activity needed to completely kill off the pathogens and bacteria.

However, if these high temperatures are reached, there is no danger of contamination, and the resulting compost can be safely used on food crops. Most composts heaps are unable to reach those temperatures. Composting toilets should only be used as a way to reduce waste in the environment, not as a fertilizer; in the case that they are used with crops, they should only use human waste for non-food crops, or with careful filtration, food based crops.

5

Role of Phytoremediation

Over the past few decades, various industrial activities discharge enormous quantities of organic and inorganic pollutants into the environment.

Controlled and uncontrolled disposal of waste, accidental and process spillage, mining and smelting of metalliferous ores, sewage sludge application to agricultural soils are responsible for the migration of contaminants into non-contaminated sites as dust or leachate and contribute towards contamination of our ecosystem.

A wide range of chemical compounds cause contamination; these include heavy metals, combustible and putriscible substances, hazardous wastes, explosives and petroleum products. Major component of inorganic contaminants are heavy metals (Adriano, 1986; Alloway, 1990), which are highly persistent (non-bio and non-thermo-degradable) in nature and thus extremely harmful to living beings because of their bioconcentration, bioaccumulation and biomagnification in food chain. Soil microorganisms can degrade organic contaminants, while metals need immobilisation or physical removal. Although many metals are essential, all metals are toxic at higher concentrations, because they cause oxidative stress by formation of free radicals.

Another reason why metals may be toxic is that they can replace essential metals in pigments or enzymes disrupting their function (Henry, 2000). Thus, metals render the land unsuitable for plant growth and destroy the biodiversity. Immobilisation or extraction of pollutants by physicochemical techniques can be expensive and is often appropriate only for small areas where rapid and complete decontamination is required. They also have an adverse effect on biological activity, soil structure and fertility and thus demand significant engineering costs. Current remediation methods applicable to soils contaminated with heavy metals are expensive, environmentally invasive, and labour intensive. A remediation technique that is of low cost, but protecting human health and the environment would be a valuable addition to current remediation methods. Salt *et al.*(1995) defined phytoremediation technique as the use of green plants to remove, contain or render harmless environmental contaminants. Plants that take up heavy metals from the soil offer an alternative and less expensive method to strip heavy metals directly from the soil as compared to other traditional treatment technologies.

Table: Cost of soil treatments

Treatment	*Cost ($/lon)*	*Additional factors/expenses*
Vitrification	75-425	Long-term monitoring
Landfilling	100-500	Transport/excavation/monitoring
Chemical treatment	100-500	Recycling of contaminants
Electrokinetics	20-200	Monitoring
Phytoextraction	5- 40	Monitoring

Environmental Biotechnology

Although natural microorganisms collectively exhibit remarkable evolutionary capabilities to adapt to a wide range of chemicals, natural evolution occurs at a relatively slow rate, particularly when the acquisition of multiple catalytic activities is necessary. In these cases, the acceleration of these events via genetic engineering is helpful, since the desirable traits can be carefully designed and controlled. The drive towards this goal represents the essence of environmental biotechnology. The International Society of Environmental Biotechnology defines environmental biotechnology as "the development, use and regulation of biological systems for remediation of contaminated environments (land, air, water), and for environment-friendly processes (green manufacturing technologies and sustainable development)". It refers to the utilisation of microorganisms to improve environmental quality. Environmental biotechnology now offer opportunities to modify organisms so that their basic biological processes are more efficient and can degrade more complex chemicals and higher volumes of waste materials. Bioremediation is addressed as one example of an environmental biotechnology, which uses living organisms to clean up contaminated soil or water.

Phytoremediation is also considered a type of bioremediation, which offers the possibility of bio-recovery of heavy metals (Yang *et al.* 2002). To breed plants having superior phytoremediation potential with high biomass production can be an alternative to improve phytoremediation. Improvement of plants by genetic engineering, i.e. by modifying characteristics like metal uptake, transport and accumulation as well as metal tolerance, opens up new possibilities for Phytoremediation.

Plants have constitutive and adaptive mechanisms for accumulating or tolerating high contaminant concentrations in their rhizospheres. Phytoremediation takes advantage of the fact that a living plant acts as a solar-driven pump, which can extract and concentrate certain heavy metals from the environment (Raskin *et.al.*,1997). This remediation method maintains the biological properties and physical structure of the soil.

Strategies of Phytoremediation

Mechanisms of metals uptake into roots and translocation to shoots.

Soluble metals can enter into the root symplast by crossing the plasma membrane of the root endodermal cells or they can enter the root apoplast through the space between cells. If the metal is translocated to aerial tissues, then it must enter the xylem. To enter the xylem, solutes must cross the Casparian strip, a waxy coating which is impermeable to solutes, unless they pass through the cells of the endodermis probably through the action of a membrane pump or channel. Once loaded into the xylem, the flow of the xylem sap will transport the metal to the leaves, where it must be loaded into the cells of the leaf, again crossing a membrane.

Once in the shoot or leaf tissues, metals can be stored in various cell types, depending on the species and the form of the metal, since it can be converted into less toxic forms (to the plant) through chemical conversion or complexation. The metal can be sequestered in several subcellular compartments (cell wall, cytosol, vacuole) or volatilised through the stomata.

Ecological studies have revealed the existence of specific plant communities, endemic floras, which have adapted on soils contaminated with elevated levels of Zn Cu, and Ni. Different ecotypes of the same species may occur in areas uncontaminated by metals. To plants endemic to metal-contaminated soils, metal tolerance is an indispensable property. In comparison, in related populations inhabiting uncontaminated areas, a continuous gradation between ecotypes with high and low tolerance usually occurs. Plants evolved several effective mechanisms for tolerating high concentrations of metals in soil. In some species, tolerance is achieved by preventing toxic metals uptake into root cells.

These plants, coined excluders, have little potential for metal extraction. Such an excluder is "Merlin," a commercial variety of red fescue *(Festuca rubra)*, used to stabilise erosion-susceptible metal-contaminated soils. A second group of plants, accumulators, does not prevent metals from entering the root. Accumulator species have evolved specific mechanisms for detoxifying high metal levels accumulated in the cells. These mechanisms allow bioaccumulation of extremely high concentration of metals. In addition, a third group of plants, termed indicators, shows poor control over metal uptake and transport processes. In these plants, the extent of metal accumulation reflects metal concentration in the rhizospheric soil. Indicator species have been used for mine prospecting to find new ore bodies (Raskin *et al.*, 1994).

Metal Detoxification in Plants

Although micronutrients such as Zn, Mn, Ni and Cu are essential for plant growth and development, high intracellular concentrations of these ions can be toxic. To deal with this potential stress, common non-accumulator plants have evolved several mechanisms to control the homeostasis of intracellular ions. Such mechanisms include regulation of ion influx (stimulation of transporter activity at low intracellular ion supply, and inhibition at high concentrations), and extrusion of intracellular ions back into the external solution.

Metal hyperaccumulator species, capable of taking up metals in the thousands of ppm, possess additional detoxification mechanisms. For example, research has shown that in *Thlaspi goesingense,* a Ni hyperacccumulator, high tolerance was due to Ni complexation by histidine which rendered the metal inactive (Kramer *et al.*, 1997; Kramer *et al.*, 1996). Sequestration in the vacuole has been suggested to be responsible for Zn tolerance in the shoots of the Zn-hyperaccumulator *Thlaspi caerulescens* (Lasat *et al.*, 1996; Lasat *et al.*, 1998). Several mechanisms have been proposed to account for Zn inactivation in the vacuole including precipitation as Zn-phytate (Van Steveninck *et al.*, 1990), and binding to low molecular weight organic acids (Tolrà *et al.*, 1996; Salt *et al.*, 1999). Complexation to low molecular weight organic compounds (<10 kD) was also shown to play a role in tolerance to Ni (Lee *et al.*, 1977). Cadmium, a potentially toxic metal, has been shown to accumulate in plants, where it is detoxified by binding to phytochelatins (Steffens, 1990; Cobbett and Goldsbrough, 1999), a family of thiol (SH)- rich peptides

(Rauser,1990). Metallothioneins (MT), identified in numerous animals and more recently, in plants and bacteria (Kägi, 1991), are also compounds (proteins) with heavy metal-binding properties (Tomsett *et al.*, 1992).

Ideal Plants for Phytoremediation

Populations of metal-tolerant, hyperaccumulating plants can be found in naturally occurring metal-rich sites (Baker and Brooks, 1989). However, these plants are not ideal for phytoremediation since they are usually small and have a low biomass production. In contrast, plants with good growth usually show low metal accumulation capability as well as low tolerance to heavy metals.

A plant suitable for phytoremediation should possess the following characteristics:

(1) Ability to accumulate the metal(s) intended to be extracted, preferably in the above ground parts (plants which do not translocate metals to the above-ground parts could be useful for phytostabilisation and landscape recreation);

(2) Tolerance to the metal concentrations accumulated;

(3) Fast growth and highly 'effective' (i.e. metal accumulating) biomass; and

(4) Easily harvestable.

Chaney *et al.* (1997) calculated that metal tolerance and hyperaccumulation would be more important to phytoremediation than high biomass production. An effective development of phytoremediation, each element must be considered separately because of its unique soil and plant chemistry.

On the other hand, metals rarely occur alone and adaptive tolerance may be needed for several metals simultaneously, even though phytoextraction of only one metal would be the goal. In some cases it might be desirable also to extract more than one metal at the same time.

Genetic Engineering to Improve Phytoremediation

To breed plants having superior phytoremediation potential with high biomass production can also be an alternative to improve phytoremediation. General plant productivity is controlled by many genes and difficult to promote by single gene insertion. Genetic engineering techniques to implant more efficient accumulator gene into other plants have been suggested by many authors (Cunningham and Ow, 1996; Brown *et.al.*, 1995; Chaney *et.al.*, 2000). Implanting more efficient accumulator genes into other plants that are taller than natural plants increases the final biomass. Zhu *et al.*, (1999) genetically engineered *Brassica juncea* to investigate rate-limiting factors for glutathione and phytochelatin production; they introduced the *Escherichia coli* -gshl- gene. The a-ECS transgenic seedlings showed increased tolerance to cadmium and had higher concentrations of Phytochelatins, a-GluCys, glutathione, and total non-protein thiols compared to wild type seedlings. The potential of success of genetic engineering can be limited because of anatomical constraints (Ow, 1996).

Several technical factors restrain the use of genetic engineering of plants for phytoremediation. One of the major factors is that there are only a few plant systems of metal resistance and/or sequestration that are sufficiently characterised to be used for this purpose. The various systems of metal resistance and accumulation are better known in microorganisms (Silver, 1996; Silver and Phung, 1996), and the first examples of their potential use in phytoremediation of metal-contaminated soils are emerging. Taking a gene from bacterium and transferring it into a plant is complicated, e.g. by the fact that the resistance is normally encoded by a large plasmid containing an operon with many genes involved in the resistance mechanism. So far, only a single gene from such an operon has been transferred (Rugh *et al.*, 1998). Generally, each metal requires specific molecular mechanism(s) for an efficient hyperaccumulation and tolerance to make plants suitable for phytoremediation. Plants accumulating metals up to several per cent of their biomass (hyperaccumulators) should be good sources for genes suitable for phytoremediation. Hundreds of such plants have been found, including *Thlaspi caerulescens* and *Brassica juncea*. Some of these hyperaccumulator plants have been used for phytoremediation without genetic engineering. However, studies on *B. juncea* showed that metal bioavailability was a major problem in Pb extraction. EDTA was thus added in soil to increase Pb availability (Blaylock *et al.*, 1997); this may, however, pose a risk to the environment. If the metal availability could be locally improved by increasing reductase activity or the amount of chelating agents, e.g. phytosiderophores (Briat and LobreAaux, 1997), without harmful effects on the environment, these hyperaccumulators might be used safely for phytoremediation.

Alteration of metal tolerance or uptake in plants through biotechnology.

Certain studies have been conducted which describes the altered metal tolerance or uptake in plants modified with foreign genes. These studies are:

Mercuric Ion Reduction and Resistance: MerA (soluble NADPH-dependent, FAD-containing disulphide oxidoreductase) is encoded by mercury resistance operon in gram-negative bacteria which includes five to six genes including mercuric ion reductase gene(merA). The enzyme coded by this gene, i.e. MerA converts toxic Hg^{2+} to less toxic Hg^0. Studies have suggested that merA gene might positively affect metal tolerance when expressed in plants. Rugh *et al.*(1996), therefore, constructed a mutagenised merA sequence (merApe9), modifying 9 per cent of the coding region and transformed it to *Arabidopsis thaliana*, The seeds germinated and the seedlings grew on the medium containing upto 100iM Hg, although the transgenic plants expressed only low levels of merAmRNA. Transgenic seedlings evolved two to three times the amount of Hg^0. This gives a good example of a successful modification of a bacterial metal tolerance gene for expression in plants. Rugh *et al.* (1998) has reported the development of transgenic yellow poplar *(Liriodendron tulipifera)* for mercury phytoremediation using merA gene (merA18) modified even further to optimise the codon usage in the plant. Transgenic plants evolved 10 times the amount of Hg^0 compared to control plants.

Mts and CD Tolerance: MTs and phytochelatins (having high percentage of cysteine sulfhydryl groups) in plants, bind and sequester heavy metal ions in very sTable complexes. Phytochelatins bind Cu and Cd with high affinity and are induced by various metals (Rauser, 1990; Ow, 1993). They may play a role in plant Cd tolerance. Howden and Cobbett (1992) have isolated *Arabidopsis* mutants with increased sensitivity to Cd while Cu tolerance was almost unchanged (Howden *et al.*, 1995 a,b). These cadl-mutants were deficient in PC synthesis and showed greatly reduced levels of PC synthase activity. MTs bind to Cu, Cd and Zn; they probably regulate intracellular Zn concentrations and detoxify normally lethal concentrations of Cd and Cu (Murphy *et al.*, 1997). MTs have the highest affinity to Cu, and MT expression correlates with Cu but not Cd tolerance in *Arabidopsis* (Murphy and Taiz, 1995). Various MT genes -mouse MTI, human MTIA (alpha domain), human MTU, Chinese hamster MTII, yeast CUP1, pea PsMTA-have been transferred to *Nicotiana* sp., *Brassica* sp. or *A. thaliana*. As a result, varying degrees of constitutively enhanced Cd tolerance have been achieved, being maximally 20-fold compared with the control. Evans *et al.*, 1992. transferred MT-like gene PsMTA of *Pisum sativum* and expressed in *A. thaliana,* more Cu (several-fold in some plants) accumulated in the roots of transformed than of control plants. The ongoing researches suggest that MT gene transfer may be useful in improving metal tolerance of plants.

Ferric Reductases and Increased Iron Uptake: Because soil contains mainly insoluble Fe(III) oxides and hydroxides, plants have adaptive mechanisms to make Fe more available for uptake. Proton extrusion is achieved by activation of an ATPase-driven proton pump. This promotes Fe(III) solubility and reduction to Fe(II) by plasma membrane-bound reductases. Fe(III) reductases are activated under Fe deficiency (Samuelsen *et al.*, 1998). FRO2 gene encoding a ferric-chelate reductase has been isolated from Fe-deficient roots of *A. thaliana*. It FRO2 gene was capable of restoring ferric-chelate reductase activity in an *Arabidopsis* mutant deficient in this enzyme (frd1) (Robinson *et al.*, 1999). The same gene also restored a mutant (frd1-1) with deficient copper-chelate reduction. Two Fe(III) reductases, FRE1 and FRE2, have been isolated from *S. cerevisiae* (Dancis *et al.*, 1990; Georgatsou and Alexandraki, 1994). Samuelsen *et al.* (1998) transferred these genes together and separately into tobacco and studied Fe(III) reductase activity and Fe accumulation under Fe-sufficient and Fe-deficient conditions in transgenic plants. In Fe- present in the growth environment, Fe(III) reduction was increased 4-fold in roots of double transformants compared to control, and also the Fe content of leaves was increased in the double transformants and FRE2 plants compared to controls and FRE1 plants. These genes may be useful for generating crops with improved nutritional quality and increased growth in Fe-deficient soils. Under suitable promoter the system may be utilised to produce plants for phytoremediation of soils polluted with more toxic metals such as Cu Though the system has not been tested in the field conditions, still these experiments gives a clear indication that applying biotechnological approaches can improve a plant's capacity to phytoremediate metal polluted sites.

Conclusion: Improvement of plants by applying biotechnological approaches opens up new possibilities for phytoremediation of metal polluted sites. However, this approach can be fully exploited only when the mechanism of the metal tolerance, accumulation and translocation are better understood. The other objectives of this approach should also be to understand the molecular mechanism that controls uptake and detoxification of heavy metals in plants, and the complex interaction between the metal, soil and plant systems.

6

Role of Rhizoremediation

"Bioremediation" is a multidisciplinary strategy or process that uses microorganisms and/or their products to degrade or detoxify pollutants from the environment (Glazer and Nikaido, 1995). With the advances in biotechnology, bioremediation has become one of the most rapidly developing fields of environmental restoration. Indigenous and enhanced microorganisms have been shown to degrade industrial solvents, polychlorinated biphenyls, explosives and many different agricultural chemicals including pesticides. Nowdays another remediation technology which is becoming popular is "Phytoremediation". Phytoremediation is the use of vegetation for *in-situ* treatment of pollutant contaminated soil, sediments and water (Juwarkar *et al.*, 2006; Newman and Reynolds, 2004,). This technology harnesses the potential of plants to degrade xenobiotics or accumulate metals to remediate contaminated sites.

A new approach which is a combination of bioremediation and phytoremediation and is more efficient in pollutant removal is "Rhizosphere Bioremediation" also known as Plant assisted bioremediation, Phytostimulated bioremediation or

Rhizoremediation (Jordahl *et al.*, 1997). It is the bioremediation process that is accelerated due to the presence of root zone. Microorganisms (yeast, fungi and/or bacteria) consume, degrade or transform organic pollutants present in the rhizosphere. Natural substances released by plant roots such as sugars, alcohols, acids etc contain organic carbon that act as nutrient source for soil microorganisms which stimulate their activity (Walker *et al.*, 2003).

Phytoremediation is a broad term that comprises several techniques used for water and soil decontamination.

Phytoextraction

It is the uptake of contaminants by plant roots and translocation within the plants. Contaminants are generally removed by harvesting plants. This technology is often applied to metal contaminated soil (Singh, 2005).

Rhizofilteration

Rhizofilteration is the precipitation or absorption of contaminants into plant roots, that are in solution surrounding the root zone, due to biotic or abiotic processes. Plant uptake, concentration and translocation might occur depending on the contaminant. Exudates from the plant roots might cause precipitation of some metals. Rhizofiltration first results in contaminant containment, in which the contaminants are immobilised or accumulated on or within the plant. Contaminants are then removed by physically removing the plant.

Phytostabilisation

It is defined as mobilisation of contaminants in soil through absorption, precipitation and/or accumulation by plant roots, within the roots zone of plants. This technology uses plants and their roots to prevent contaminants migration via wind and water erosion, leaching and soil dispersion. Phytostabilisation occurs through root zone microbiology and chemistry and alteration of the soil environment or contaminant chemistry. Soil pH may be changed by plant roots exudates or through the production of CO_2. It can change metal solubility and mobility or influence the degradation of organic compounds.

The term phytolignification has been used to refer to a form of phytostabilisation in which organic compounds are incorporated into plant lignin (Cunningham *et al.*, 1995).

Rhizodegradation

It is the breakdown of organic contaminant in soil through microbial activity that is enhanced by the presence of root zone. Rhizodegradation is also known as plant assisted degradation, plant assisted bioremediation, plant aided *in-situ* biodegradation, and enhance rhizosphere biodegradation.

Phytodegradation

Also known as phyotransformation. It is the breakdown of contaminants taken up by plants through metabolic process within the plants or the break down of contaminants external to the plant through the effect of compounds like enzymes, produced by the plants.

Phytovolatilisation

It is the uptake and transpiration of a contaminant by plants, with release of the contaminant or its modified form to the atmosphere from the plant through contaminant uptake, plant metabolism, and plant transpiration. Phytodegradation is related phytoremediation process that can occur along with phytovolatilisation.

Hydraulic Control

Hydraulic control is the use of vegetation to remove groundwater through uptake and consumption in order to contain or control the migration of contaminants. Hydraulic control is also known as phytohydraulics or hydraulic plume control.

Rhizoremediation

The remediation process in which the contaminants are transformed by microbes in the rhizosphere is referred to as rhizodegradation/rhizosphere bioremediation/enhanced rhizosphere biodegradation. Rhizodegradation is the enhancement of naturally-occurring biodegradation in soil through the influence of plant roots, and ideally leads to destruction or detoxification of an organic contaminant (Singh and Jain, 2003). Other terms have been used by some authors as synonyms for rhizodegradation, such as enhanced rhizosphere biodegradation. Because of the involvement of plants and their active role in this remediation technology most authors consider it as one of the subtypes of phytoremediation technologies.

Rhizosphere

The term rhizosphere refers to the environment influenced by roots of vascular plants in which elevated bacterial activity is observed. The general rhizosphere effect is well known and has been described as the zone of soil under the direct influence of plant roots, which visually extends a few millimetres from the root surface and is a dynamic environment for microorganisms.

The rhizosphere provides a complex and dynamic microenvironment where bacteria, fungi and actinomycetes along with other microorganisms, in association with roots, form unique communities that have considerable potential for detoxification hazardous organic recalcitrant compounds. The stimulatory rhizosphere effect was first described by Hiltner 1904 (Kuiper *et al.*, 2003). The main reason for improved degradation in rhizosphere is presumably the increase in the number and metabolic activities of microbes. The exudation of nutrients by plant roots creates a nutrient rich environment in which microbial activity is stimulated. Plant roots exudates contains sugars, organic acids and amino acids as major components. In addition, the mucigel secreted by root cells, lost root cap cells, or the decay of complete roots provide nutrients of the rhizospheric microorganisms (Vancura and Hovadik, 1965).

Table: Summarises mechanisms, application media, process goals, types of contaminants, plants used, and present status of various phytoremediation technologies EPA/600/R-99/107 (2000)

Mechanism	*Process Goal*	*Media*	*Contaminants*	*Plants*	*Status*
Phytoextraction	Contaminant extraction and capture	Soil, sediment, and sludges	Metals: Ag, Cd, Co, Cr, Cu, Hg, Mn, Mo, Ni, Pb, Zn; Radionuclides: 90Sr, 137Cs, 239Pu, 238, 234U	Indian mustard, pennycress, alyssum, sunflowers, hybrid poplars	Laboratory, pilot, field applications
Rhizofiltration	Contaminant extraction And capture	Groundwater, surface water	Metals, radionuclides	Sunflowers, Indian mustard, water hyacinth	Laboratory and pilot scale
Phytostabilisation	Contaminant containment	Soil, sediment, sludqes	As, Cd, Cr, Cu, Hs, Pb,Zn	Indian mustard, hybrid poplars. poplars, grasses	Field application
Rhizodegradation	Contaminant destruction	Soil, sediment, sludges, ground water	Organic compounds (TPH, PAHs, pesticides chlorinated solvents, PCBs)	Red mulberry, grasses, hybrid poplar, cattail, rice	Field application
Phytodegradation	Contaminant destruction	Soil, sediment, sludges, groundwater surface water	Organic compounds, chlorinated solvents, phenols, herbicides, munitions	Algae, stonewort, hybrid poplar, black willow, bald cypress	Field demonstration
Phytovolatilisation	Contaminant extraction from media and release to air	Groundwater, soil, sediment, sludges	Chlorinated solvents, some inorganics (Se, Hg, and As)	Poplars, alfalfa black locust, Indian mustard	Laboratory and field application
Hydraulic control (plume control)	Contaminant egradation or containment	Groundwater, surface water	Water-soluble organics and inorganics	Hybrid poplar, cottonwood, willow	Field application
Vegetative cover (evapotranspiration cover)	Containment, erosion control	Soil, sludge, sediments	Organic and inorganic compounds	Poplars, grasses	Field application
Riparian corridors (non-point source control)	Contaminant destruction	Surface water, groundwater	Water-soluble organics	Poplars	Field demonstration

Many rhizospheric microorganisms are known to possess either plant growth promoting effect or plant growth inhibiting properties. Microbial diversity in the rhizosphere depends on various factors like plant species, composition of root exudates, root type, plant age, soil type and soil history.

Since the roots supply nutrients, viz. amino acids, carbohydrates, and organic acids (Aprill and Sims, 1990), so no exogenous carbon source is needed to add, and they also supply bacteria with cofactors required for the activation of bacterial enzymes involved in the pollutant degradation pathway.

Root Exudates

In addition to accumulating biologically active chemicals, plant roots continuously produce and secrete compounds into the rhizosphere.

Table: Common bacterial genera in soil

Bacteria	*Function*
Azospirillum	N-fixation in tropical grasses
Pseudomonaceae	Decompose organic matter
Azotobacteriaceae	Free N fixers
Rhizobiaceae	
Rhizobium	Smbiotic N fixers,produce root nodules in legumes
Agarobacterium	Produces root infections
Enterobacter(Aerobacter)	Decomposition of organic matter
Nitrobacteriaceae	Oxidise reduced form of inorganic Nitrogen
Endospore formers	
Bacillus	Some spp. Fix Nitrogen, produce a number of extracellular enzymes which hydrolyse complex organic material
Closthdium	Ferment cellulose, starch, pectin and sugars
Actinomycetes (Largest and heterogenous group)	
Arthobacter	
Nocardia	Use variety of organic compounds and produce antibiotics
Actinomyces	
Streptomyces	
Frankia	Fix nitrogen in non- legumes

Root exudation includes the secretion of ions, free oxygen and water, enzymes, mucilage, and a diverse array of carbon-containing primary and secondary metabolites. Plant exudates include sugars, amino acids, organic acids, fatty acids, sterols, growth factors, nucleotides, flavanones, enzymes, and other compounds (Shimp *et al.*, 1993).

Root exudation can be broadly divided into two active processes:

- *Root Excretion:* It involves gradient dependent output of waste materials with unknown functions.
- *Secretion:* involves exudation of compounds with known functions, such as lubrication and defence.

Roots release compounds via at least two potential mechanisms:

- Root exudates are transported across the cellular membrane and secreted into the surrounding rhizosphere.
- Plant products are also released from root border cells and root border-like cells, which separate from roots as they grow.

Root exudates are divided into:

- Low-molecular weight compounds such as amino acids, organic acids, sugars, phenolics, and other secondary metabolites account for much of the diversity of root exudates.
- High molecular weight exudates, such as mucilage (polysaccharides) and proteins, are less diverse but often compose a larger proportion of the root exudates by mass.

Root exudation clearly represents a significant carbon cost to the plant, and the magnitude of photosynthesis secreted as root exudates varies with the type of soil, age, and physiological state of the plant, and nutrient availability. Although, the functions of most root exudates have not been determined, several compounds present in root exudates play important roles in biological processes.

Rhizoremediation is an attractive process since plant roots provide a large surface area for a significant population of bacteria and transport the root-colonising, remediating microorganism to pollutants upto 10 to 15 m deep in the soil (Kingsley *et al.*, 1994). Consequently, rhizoremediation appears to be an aesthetically pleasing, low-cost, minimal-maintenance, *in-situ* treatment for pollutants in surface soils.

Number of processes are involved in detoxification of pollutants including degradation, mineralisation and polymerisation. Apart from rhizospheric micro consortia other factors, which influence the detoxification process are: properties of host plant, soil properties and environmental conditions. Complete understanding of the interactions among plants, rhizosphere microbial communities and organic toxicants is essential for the successful utilisation of rhizosphere technology to remediate contaminated soils.

Role of Mycorrhizal Associations in Rhizoremediation

Among the microorganisms that affect rhizosphere processes, symbiotic fungi forming mycorrhizae induce a series of changes in plant physiology, nutrient availability and microbial composition that may determine the outcome of a phytoremediation attempt. Beyond the rhizosphere, mycorrhizal hyphae act as the roots of the roots, and may thus extend the rhizosphere into the bulk soil by creating a new interface of soil-plant interactions: the hyphosphere.

Mycorrhizae can be defined as a symbiotic association between fungi (myco) and plant roots (rhizae). Studies with mycorrhizal fungus have focused on their ability to provide the host plant with nutrients in nutrient deficient soil and have ignored their other roles in soil ecosystem. Although mycorrhizal fungi are well known as saprophytic microorganism there is possibility that they may retain the ability to degrade a large assortment of compounds available in the soil, including organic pollutants. Mycorrhizal fungi are usually classified as belonging to one of four major types: the ectomycorrhizae, the vesicular arbuscular mycorrhizae, the ericaceous mycorrhizae, and the orchid mycorrhizae.

An important benefit of mycorrhizal infection has been observed in sites with heavy metal contamination, such as mine spoils and at sites containing high concentration of other toxic metals. Plants with mycorrhizal associations appear to be protected from the phytotoxic effect of such heavy metals. It is believed that the mycorrhizal fungi present in the contaminated sites have been selected over time with greater resistance to heavy metals (Juwarkar *et al.*, 2002). Heavy metals are believed to be bound by carboxyl groups in the hemicelluloses of the interfacial matrices between the host cells and the fungus. An ericoid mycorrhizal fungus, *Hymenoscyphus ericae,* was able to metabolise some phytotoxic compounds *in vitro* (Leake, 1987). This ability of the fungus provides the host plant with added protection against toxins and allows the plants to grow in areas, otherwise hazardous to plant.

Donelly *et al.*, (1993) found that many of the fungi were capable of degrading 2,4,D and atrazine. The two ericoid mycorrhizal fungi, *Hymenoscyphus ericae* and *Oidiodendron griseum* and two ectomycorrhizal fungi *Rhizopogon vulgaris* and *Gautieria crispa* were able to degrade significant amount of atrazine. In most case degradation consisted of breakdown of compounds and incorporation of carbon into tissues rather than mineralisation. Also the degradation of other herbicides like hexazinone and chloropropham, by Mycorrhizal fungi has been studied (Robideaux *et al.*, 1991). Many mycorrhizal fungi are capable of degrading PCBs. Out of the total 21 fungal species tested 14 were able to metabolise PCBs by at least 20 per cent. The ectomycorrhizal fungi *Radiigera atrogleba* and *Hysterangium gardneri* were able to degrade over 80 per cent of 2,2'-dichlorobiphenyl.

Although, all the experiments conducted to study the

degradation of organic pollutants by mycorrhizal fungi are *in vitro* without host plants, but the results well establish the fact that this group of fungi have the ability to degrade a range of xenobiotics. *In vitro* studies with intact symbiosis are now in progress. Both enzymatic and physiological studies completed with mycorrhizal fungi indicate that these fungi have the potential to play an important role in bioremediation of hazardous compounds in soil. The natural, sustained occurrence of abundant amounts of these fungi in association with plants make this plant-microbe system an ecologically sound approach to long term bioremediation and ecological restoration of contaminated sites.

Role of Phytochelatins, Phytosiderophores in Rhizoremediation

Some root exudates that act as metal chelators in the rhizosphere can increase the availability of metallic soil micronutrients, including iron, manganese, copper, and zinc. Metal chelators form complexes with soil metals, thus releasing metals that are bound to soil particles and increasing metal solubility and mobility. Immobilisation of heavy metals by cysteine-rich peptides is a major mechanism employed by plants for counteracting heavy metal toxicity. Phytochelatins (PC), have been shown to bind heavy metals with high affinity (Dakora and Phillips 2002).

Because many plant species secrete phytosiderophores (chelating agents) to solubilise soil Fe, and accumulate the intact chelate into root cells, it is suggested that transgenic plants could be developed to secrete metal selective ligands into the rhizosphere which could specifically solubilise elements of phytoremediation interest. Although this approach holds promise, phytosiderophores obtain their specificity not by chelation specifically only of Fe in soils, but from their uptake of nearly only Fe-phytosiderophores by a membrane carrier (Yehuda, 1996).

Extensive progress has recently been achieved in identifying genes and proteins involved in uptake of Fe by yeast and plants (Yi *et al.*, 1996). A fundamental understanding of both uptake and translocation processes in normal plants and metal hyperaccumulators, regulatory control of these activities, and the use of tissue specific promoters offers great promise that use of molecular biology tools can give scientists the ability to develop effective and economic phytoremediation plants for soil metals.

Rhizodegradation

It is the term used to describe the process of enhanced microbial degradation/mineralisation of recalcitrant compounds in root zone, i.e. Rhizpsphere.

Organic contaminants in soil are often broken down into daughter products or completely mineralised to inorganic products such as carbon dioxide and water by naturally occurring bacteria, fungi, and actinomycetes. The presence of plant roots increases the size and variety of microbial populations in the soil surrounding roots (the rhizosphere) or in mycorrhizae (associations of fungi and plant roots). Significantly higher populations of total heterotrophs, denitrifiers, pseudomonads, BTEX (benzene, toluene, xylenes) degraders, and atrazine degraders were found in rhizosphere soil around hybrid poplar trees in a field plot *(Populus deltoides* × nigra DN-34, Imperial Carolina) than in non-rhizosphere soil (Jordahl *et al.*, 1997). The increased microbial populations is due to stimulation by plant exudates, compounds produced and released from plant roots.

Either the compounds present in root exudates like sugars, amino acids, organic acids, fatty acids, sterols, growth factors, nucleotides, flavanones, enzymes, or their degradation products can stimulate cometabolism of contaminants in the rhizosphere. Plants also release exudates in the soil environment that may help to stimulate the degradation of organic chemicals by inducing enzyme systems of existing bacterial populations, stimulating growth of new species that are able to degrade the wastes. This effect has not been addressed in most phytoremediation research. Plants may also enhance biodegradation by other mechanisms such as breaking apart clods and transporting atmospheric oxygen to root zone. Rhizodegradation occurs primarily in soil, although stimulation of microbial activity in the root zone of aquatic plants could potentially occur. Stimulation of soil microbes by plant root exudates can also result in alteration of the geochemical conditions in the soil, such as pH, which may result in changes in the transport of inorganic contaminants.

Plants and plant roots can also affect the water content, water and nutrient transport, aeration, structure, temperature, pH, or other parameters in the soil, often creating more favourable environments for soil microorganisms, regardless of the production of exudates. Thus a plant can be considered as a solar driven biological pump and treatment system, attracting water with its roots system, accumulating water soluble pollutants in the rhizosphere and concluding with the degradation or translocation of the pollutant (Erickson, 1997).

The fate of organic contaminants in the environment is associated with both abiotic and biotic processes, including chemical oxidation, bioaccumulation and microbial transformation. The specific interactions of a pollutant with soil, water and plants will vary depending on the chemical properties of the contaminant, the physiological properties of the introduced plant species and the contaminated medium. Organic compounds can be translocated to other plant tissues (Salt *et al.* 1998) and subsequently volatilised, they may undergo partial or complete degradation or they may be transformed to less phytotoxic compounds and bound in plant tissues. Collectively, these properties determine whether a contaminant is subjected to phytoextraction, phytodegradation, phytovolatilisation or rhizodegradation.

Studies by Narayanan *et al.*, (1995) reflected the possibility that transpiration due to alfalfa plants drew methane from a saturated methanogenic zone up into the vadose zone where the methane was used by methanotrophs that cometabolically degraded trichloroethylene (TCE). Lin and Mendelssohn (1998) indicated that the salt marsh grasses *Spartina alterniflora* and *S. patens* could potentially increase subsurface aerobic biodegradation of spilled oil by transporting oxygen to their roots. Rizodegradation is responsible for the enhanced removal of total petroleum hydrocarbons from soil by deep-rooted trees and other annual species. A wide range of organic contaminants are candidates for rhizodegradation, such as petroleum hydrocarbons, PAHs, pesticides, chlorinated solvents, PCP, polychlorinated biphenyls (PCBs), and surfactants. Higher populations of benzene-, toluene-, and o-xylene-degrading bacteria were found in soil from the rhizosphere of poplar trees than in non-rhizosphere soil.

Jordahl *et al.,* (1997) observed total petroleum hydrocarbon (TPH) disappearance at several field sites contaminated with crude oil, diesel fuel, or petroleum refinery wastes, at initial petroleum hydrocarbon contents of 1,700 to 16,000 mg/kg TPH.

Advantages

Rhizodegradation has the following advantages:

- Degradation process is *in-situ*.
- Translocation of the compound to the plant or atmosphere is less likely than with other phytoremediation technologies since degradation occurs at the source of the contamination.
- Complete mineralisation of the contaminant can occur.
- Low installation and maintenance cost as compared to other remedial options.

Disadvantages

Rhizodegradation has the following disadvantages:

- Development of an extensive root zone is required which takes substantial time.
- Root depth can be limited due to the physical structure or moisture conditions of the soil.

- Plant uptake can occur for many of the contaminants. Laboratory and field studies need to account for other loss and phytoremediation mechanisms that might complicate the interpretation of rhizodegradation. For example, if plant uptake occurs, phytodegradation or phytovolatilisation could occur in addition to rhizodegradation.
- The exudates might stimulate microorganisms that are not degraders, at the expense of degraders.
- Organic matter from the plants may be used as a carbon source instead of the contaminant, which could decrease the amount of contaminant biodegradation. In laboratory sediment columns, debris from the salt marsh plant *Spartina alterniflora* decreased the amount of oil biodegradation. This could have been due to competition for limited oxygen and nutrients between the indigenous oil-degrading microorganisms and the microorganisms degrading plant organic matter (Narayanan *et al.* 1995).

Applications

The following contaminants are amenable to rhizodegradation:

Total Petroleum Hydrocarbons (TPH)

- Several field sites contaminated with crude oil, diesel, and other petroleum products were studied for phytoremediation by examining TPH disappearance.
- Rhizodegradation and humification were the most important disappearance mechanisms, with little plant uptake occurring.

Polycyclic Aromatic Hydrocarbons

General pathway for ring cleavage of polycyclic aromatic hydrocarbons (PAHs) followed by microorganisms in soil is shown in the following figure:

Benzene, Toluene, Ethyl benzene and Xylenes (BTEX)

- Soil from the rhizosphere of poplar trees had higher

populations of benzene-toluene, and o-xylene-degrading bacteria than did non- rhizosphere soil. Root exudates contained readily biodegradable organic macromolecules (Jordahl *et al.*, 1997).

- Chrysene, benzo anthracene, benzo pyrene, and dibenzo anthracene had greater disappearance in vegetated soil than in non- vegetated soil (Aprill and Sims, 1990).
- Pyrene was mineralised at a greater rate in a planted system than in an unplanted system (Ferro *et al.*, 1994a).

Pyrene at 150 mg/kg was used in an experiment with crested wheatgrass (Ferro *et al.*, 1994a).

- Anthracene and pyrene at 100 mg/kg were used in a study with grasses and a legume (Reilley *et al.*, 1996).
- 10 mg/kg PAH (chrysene, benzo anthracene, benzo pyrene, dibenzo anthracene) had greater disappearance in vegetated soil than in non-vegetated soil (Reilley *et al.*, 1996).

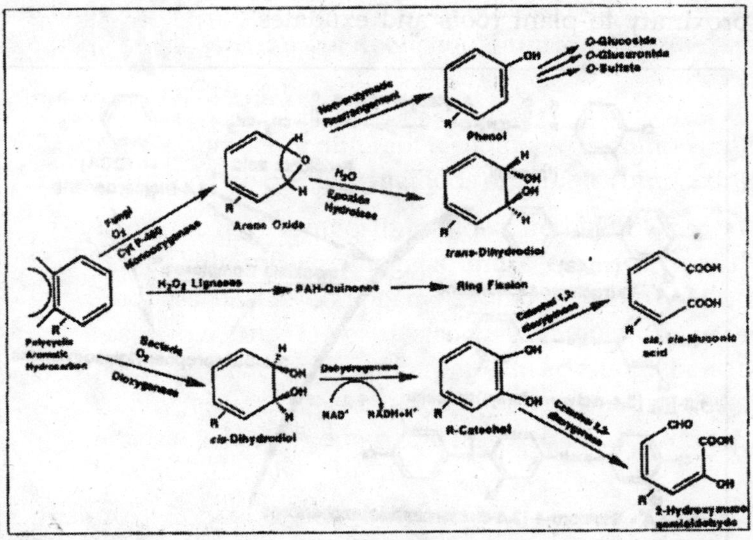

Figure: Generalised eukaryotic and prokaryotic pathway for oxidative metabolism of PAHs (Cerniglia, 1992)

Pesticides

Atrazine, Metolachlor, and Trifluralin Herbicides

Soil from the rhizosphere had increased degradation rates compared to non-rhizosphere soil. The experiments were conducted in the absence of plants to minimise effects of root uptake (Anderson *et al.*, 1994).

Parathion and Diazinon Organophosphate Insecticides

Mineralisation rates of the radiolabelled compounds were higher in rhizosphere soil (soil with roots) than in non-rhizosphere soil (soil without roots). Diazinon mineralisation in soil without roots did not increase when an exudate solution was added, but parathion mineralisation did increase (Hsu and Bartha, 1979).

Propanil Herbicide

An increased number of gram negative bacteria were found in rhizosphere soil (Hoagland *et al.*, 1994). It was hypothesised that the best propanil degraders would benefit from the proximity to plant roots and exudates.

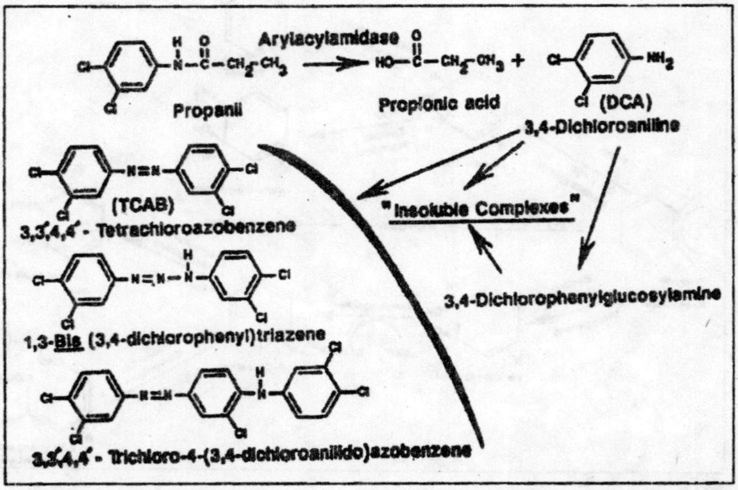

Figure: Schematic representation of propanil metabolism and/or degradation by microorganisms (Anderson and Coats, 1994)

2,4-D Herbicide

Microorganisms capable of degrading 2,4-D occurred in elevated numbers in the rhizosphere of sugar cane, compared to non-rhizosphere soil. The rate constants for 2,4-D biodegradation were higher in rhizosphere soil than in non-rhizosphere soil (Boyle and Shann, 1995).

T Herbicide

The rate constants for 2,4,5-T biodegradation were higher in rhizosphere soil than in non-rhizosphere soil (Boyle and Shann, 1995).

Chlorinated Solvents Tetrachloroethelyene (TCE)

Greater TCE mineralisation was measured in vegetated soil as compared to non-vegetated soil. TCE and 1,1,1-trichloreethane (TCA) dissipation was possibly aided by rhizosphere biodegradation enhanced by the plant roots TCE at 100 and 200 g/L in groundwater was used in a soil and groundwater system.

Pentachlorophenol (PCP)

PCP was mineralised at a greater rate in a planted system than in an unplanted system (Ferro *et al.*, 1994b). 100 mg PCP/kg soil was used in an experiment with hycrest crested wheatgrass *(Agropyron desertorum)*

Polychlorinated Biphenyls (PCBs)

Compounds (such as flavonoids and coumarins) found in leachate from roots of specific plants stimulated the growth of PCB-degrading bacteria (Gilbert and Crowley, 1997).

Surfactants

Linear alkylbenzene sulphonate (LAS) and linear alcoholethoxylate (LAE) had greater mineralisation rates in the presence of root microorganisms than in non-rhizosphere sediments LAS and LAE at 1 mg/L had greater mineralisation rates in the presence of root microorganisms than in non-rhizosphere sediments (Federle and Schwab, 1989).

Explosives

Within one week, the dissolved TNT concentrations in flooded soil decreased from 128 ppm to 10 ppm in the presence of the aquatic plant parrot feather *(Myriophyllum aquaticum)*, which produces nitroreductase enzyme that can partially degrade TNT (Schnoor *et al.*, 1995).

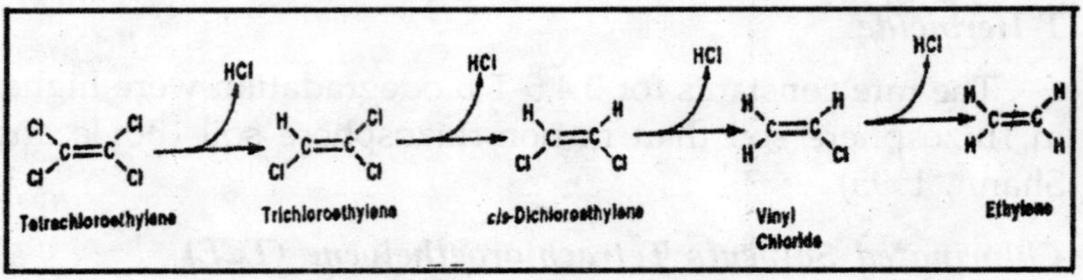

Figure: Reductive dechlorination of tetrachloroethylene to ethylene as mediated by bacterial transition metal coenzyme F430 and vitamin B12 (Aust, 1990)

The nitroreductase enzyme has also been identified in a variety of algae, aquatic plants, and trees. Hybrid poplar trees metabolised TNT to 4-amino-2,6-dinitrotoluene (4-ADNT), 2-amino-4,6-dinitrotoluene (2-ADNT), and other unidentified compounds in laboratory hydroponic and soil experiments (Thompson *et al.*, 1998).

Because the rhizosphere extends only about 1 mm from the root and initially the volume of soil occupied by roots is a small fraction of the total soil volume, the soil volume initially affected by the rhizosphere is limited. With time, however, new roots will penetrate more of the soil volume and other roots will decompose, resulting in additional exudates to the rhizosphere. Thus, the extent of rhizodegradation will increase with time and with additional root growth. The effect of rhizodegradation might extend slightly deeper than the root zone. If the exudates are water soluble, not strongly sorbed, and not quickly degraded, they may move deeper into the soil. Contaminated groundwater can be affected if it is within the influence of roots. Plants that produce exudates that have been shown to stimulate growth of degrading microorganisms or stimulate cometabolism will be of more benefit than plants without such directly useful exudates. The type, amount, and effectiveness of exudates and enzymes produced by a plant's roots will vary between species and even within subspecies or varieties of one species.

Although there are a number of reports on enhanced degradation/mineralisation of recalcitrant xenobiotics in the rhizosphere, the exact mechanism is still not clear. Some of the suggested reasons for rhizosphere-enhanced biodegradation are as follows:

- As a result of rhizodeposition, stimulation of high densities of active and diverse heterotrophic microorganisms on root surfaces, a situation which will facilitate complex degradative pathways through activities of consortia and conjugative horizontal transfer of catabolic genes carried on mobile genetic elements (Crowley, *et al.*, 1997).
- Another suggested reason for rhizosphere-enhanced xenobiotic biodegradation is that components of the rhizodeposits (exudates, root materials, or their decomposition products) may be structural analogs for the pollutant in question. It follows that the presence of a pollutant analog in the rhizosphere may select for pollutant-degrading microorganisms capable of directly using the analog for growth, act as a cooxidised substrate during cometabolism, or serve as an effector molecule in the induction of xenobiotic catabolic pathways (Haby and Crowley, 1996).
- Furthermore, some investigators (Siciliano *et al.*, 2001) hold the view that the rhizodeposition of compounds stimulating xenobiotic biodegradation may not be merely a passive process that fortuitously results in enhanced biodegradation. On the contrary, it is suggested that the plant, on sensing the presence of a xenobiotic in its rhizosphere, actively alters the composition of its rhizodeposits to stimulate xenobiotic degradation and that this response is an extension of

the plant's defences against natural allelopathic chemicals. Plants Applied for Rhizodegradation

Rhizofiltration

Rhizofiltration (also known as phytofiltration) is the removal of contaminants in surface water, waste water, or extracted ground water, by plant roots through adsorption or precipitation. The root environment or root exudates may produce biogeochemical conditions that result in precipitation of contaminants onto the roots. The contaminant may remain on the root, within the root, or be taken up and translocated into other portions of the plant, depending on the contaminant, its concentration, and the plant species.

Rhizofiltration and phytoextraction are similar in that they each result in accumulation of the contaminant in or on the plant, however, in rhizofiltration this accumulation can occur in the roots or in the portion of the plant above water, whereas for effective phytoextraction the accumulation occurs aboveground, not in the roots.

Table: Summarises the plant species mostly used for degradation of different types of pollutants

Plant Species	*Contaminants*	*References*
Red mulberry *(Morus rubra)* Crabapple *(Maius fusca)* Osaqe orange *(Madura pomifera)*	PCB	Fletcher and Hegde, 1995
Spearmint *(Mentha spicata)*	PCB	Gilbert and Crowley, 1997
Alfalfa *(Medicago sativa)* *(Lespedeza cuneata)*	TCE and TCA	Narayanan *et al.*, 1995
Loblolly pine *(Pinus taeda)* Soybean *(Glycine max)*	TCE	Anderson and Walton, 1995
Rye grass St. Augustine grass	TPH	Schwab, 1998
Big bluestem *(Andropogon gerardi)* Little bluestem *(Schizachyrium scoparius)*, Indian grass *(Sorghastwm nutans)*, Switch grass *(Panicum virgatum)*, Canada wild rye *(Elymus canadensis)*, Western wheatgrass *(Agropyron smithii)*, Side oats grama *(Bouteloua curtipendula)*, Blue grama *(Bouteloua gracilis)*	PAH	Aprill and Sims, 1990

Contd...

Plant Species	**Contaminants**	**References**
Fescue *(Festuca arundinacea Schreb)* Sudan grass *(Sorghum vulgare)* Switch grass *(Panicum virgatum),* Alfalfa *(Medicaqo sativa)*	PAH	Reilley *et al.*, 1996
Wheatgrass *(Agropyron desertorum)*	PCP and Pyrene	Ferro *et al.* 1994a, 1994b

Table: Plant Species Applied for Rhizodegradation

Plant Species	**Contaminants**	**References**
Hybrid poplar tree *(Populus deltoides)*	BTEX, Atrazine	Jordahle *et al.*, 1997
Fescues (*Festuca ovina var duriuscula, Festuca arundinacea, Festuca rubra)* Wheat grasses *(Agropyron smithii)* Bluestem *(Andropogon scoparius),* Indian grass *(Sorghastrum nutans),* Switch grass *(Panicum virgatum)*	PAH and PCP	Jordahlera *et al.*, 1997
Bush bean *(Phaseolus vulgaris)*	Parathion and Diazinon	Hsu and Bartha, 1979
Rice *(Oryza sativa)*	Propanil	Hoagland *et al.*, 1994
Kochia sp.	Herbicides	Anderson *et al.*, 1994
Cattail *(Typha latifolia)*	LAS and LAE	Federle and Schwab, 1989

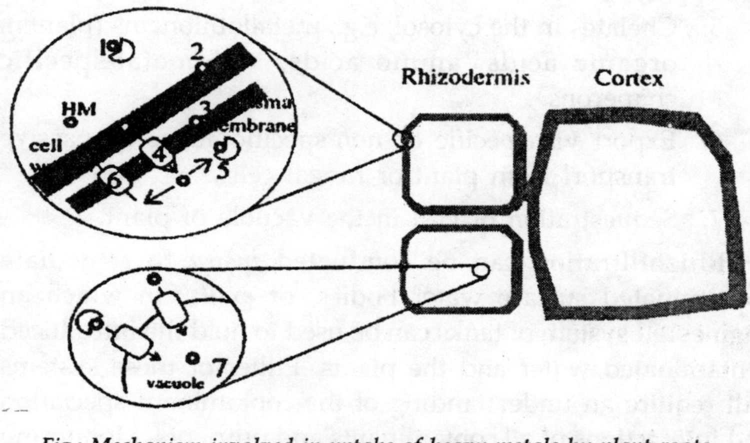

Fig.: Mechanism involved in uptake of heavy metals by plant roots (Gohre and Paszkowski, 2006)

In addition, rhizofiltration differs from phytoextraction in that the contaminant is initially in water, rather than in soil. Rhizofiltration is a contaminant removal process, in which contaminant removal from the site is accomplished by harvesting the roots and, if necessary, the above-water portion of the plant, followed by proper disposal of the contaminated plant mass. Thus, rhizofiltration differs from phytostabilisation occurring in soil, in which the contaminant remains in the root zone. Rhizofiltration is generally applicable to treating large volumes of water with low contaminant concentrations (in the ppb range). It has primarily been applied to metals (Pb, Cd, Cu, Fe, Ni, Mn, Zn, Cr(VI) and radionuclides 90Sr, 137Cs, 238U, 236U (Salt *et al., 1998*, Dushenkov *et al.*, 1997).

Heavy Metal Detoxification Mechanisms in Plants

1. Chelating agents are secreted that bind metals in the soil, e.g. histidine and organic acids from the plant.
2. Binding of HM to cell wall components in plants.
3. The plasma membrane as a living, selective barrier in plants.
4. Specific and non-specific metal transporters and pores in the plasma membrane of plants (active and passive import).
5. Chelates in the cytosol, e.g., metallothioneins (plants), organic acids, amino acids, and metal-specific chaperons.
6. Export via specific or non-specific active or passive transport from plant or fungal cells.
7. Sequestration of HM in the vacuole of plant.

Rhizofiltration can be conducted *insitu* to remediate contaminated surface water bodies, or *exsitu,* in which an engineered system of tanks can be used to hold the introduced contaminated water and the plants. Either of these systems will require an understanding of the contaminant speciation and interactions of all contaminants and nutrients. Monitoring and possible modification of the water pH, or of the flow rate and contaminant concentration of influent water, is necessary. Predictions of metal immobilisation and uptake from laboratory studies and greenhouse studies might not be achievable in the field. However, in an engineered *exsitu* system, the ability to control conditions may allow results to approach those predicted in the laboratory. Effluent from engineered flow-through rhizofiltration systems will need to meet relevant discharge limits. Proper disposal of the contaminated plant biomass is required. Applications of rhizofiltration are currently at the pilot-scale stage. Phytotech tested a pilot-scale rhizofiltration system in a greenhouse at a Department of Energy uranium-processing facility in Ashtabula, Ohio (Dushenkov *et al.*, 1997).

This engineered ex-situ system used sunflowers to remove uranium from contaminated ground water and/or process water. Phytotech also conducted a small-scale field test of rhizofiltration to remove radionuclides from a small pond near the Chernobyl reactor, Ukraine. Sunflowers were grown for four to eight weeks in a floating raft on a pond, and bioaccumulation results indicated that sunflowers could remove 137Cs and 90Sr from the pond.

The basic principles of detoxification mechanisms include the extracellular Heavy Metal (HM)-chelation by root exudates and/or binding of HM to the rhizodermal cell walls uptake of HM avoiding. Active plant efflux systems control cytosolic concentrations of HM. Intracellularly the plant cell produces chelating agents such as phytochelatins and metallothioneins, which have high-affinity HM binding properties. The resulting complex can finally be exported from the cytoplasm across the tonoplast and become sequestered inside the vacuole (Hall 2002). Also other organelles are involved in storage. In plants Fe is stored bound to ferritin inside chloroplasts (Briat and Lobreaux, 1997).

This technology can be used to treat extracted groundwater, surface water, and waste water. Rhizofiltration is generally applicable to low-concentration, high-water-content conditions.

This technology does not work well with soil, sediments, or sludges because the contaminant needs to be in solution in order to be sorbed to the plant system.

Advantages

Rhizofiltration offers following advantages:

- Either terrestrial or aquatic plants can be used. Although terrestrial plants require support, such as a floating platform, they generally remove more contaminants than aquatic plants.
- This system can be either *in-situ* (floating rafts on ponds) or *exsitu* (an engineered tank system).
- An *exsitu* system can be placed anywhere because the treatment does not have to be at the original location of contamination.

Disadvantages

Rhizofiltration has the following disadvantages:

- A well-engineered system is required to control influent concentration and flow rate.
- Periodic harvesting and plant disposal are required.
- The pH of the influent solution may have to be continually adjusted to obtain optimum metals uptake.
- The chemical speciation and interaction of all species in the influent have to be understood and accounted for.
- The plants (especially terrestrial plants) may have to be grown in a greenhouse or nursery and then placed in the rhizofiltration system.
- Metal immobilisation and uptake results from laboratory and greenhouse studies might not be achievable in the field.

Applications

Rhizofilteration can be used for removal of metals from aqueous medium:

Lead

Pb^{2+} at a solution concentration of 2 mg/L, was accumulated in Indian mustard roots with a bioaccumulation coefficient of 563 after 24 hours. Pb^{2+} at solution concentrations of 35, 70, 150, 300, and 500 mg/L was accumulated in Indian mustard roots, although root adsorption of Pb saturated at 92 to 114 mg Pb/g DW root. Pb disappeared from the 300- and 500-mg/L solutions due to precipitation of lead phosphate. Pb absorption by roots was found to be rapid, although the amount of time required to remove 50 per cent of the Pb from solution increased as the Pb concentration increased (Dushenkov *et al.*, 1995). Pb was accumulated in the roots of Indian mustard *(Brassica juncea)* in water concentrations of approximately 20 to 2,000 g/L, with bioaccumulation coefficients of 500 to 2,000 (Salt *et al.*, 1997). Pb at concentrations of 1 to 16 mg/L was accumulated by water milfoil *(Myriophyllum spicatum)* with a minimum residual concentration below 0.004 mg/L (Wang *et al.*, 1996).

Cadmium

Cd^{2+} (2 mg/L) was accumulated in Indian mustard roots with a bioaccumulation coefficient of 134 after 24 hours (Dushenkov *et al.*, 1995). Cd was accumulated by the roots of Indian mustard *(Brassica juncea)* in water concentrations of about 20 to 2,000 g/L, with bioaccumulation coefficients of 500 to 2,000. The seedlings removed 40 to 50 per cent of the Cd within 24 hours. The Cd went from 20 g/L to 9 g/L within 24 hours. After 45 hours, the Cd reached 1.4 per cent in the roots and 0.45 per cent in the shoots. Cd saturation was reached in the roots in 12 hours and in the shoots in 45 hours. Cd at concentrations of 1 to 16 mg/L was accumulated by water milfoil *(Myriophyllum spicatum)* with a minimum residual concentration of approximately 0.01 mg/L (Wang *et al.*, 1996).

Copper

Cu^{2+} (6 mg/L) was accumulated in Indian mustard roots with a bioaccumulation coefficient of 490 after 24 hours (Dushenkov *et al.*, 1995). Cu at concentrations of 1 to 16 mg/L was accumulated by water milfoil *(Myriophyllum spicatum)* with a minimum residual concentration of approximately 0.01 mg/L (Wang *et al.*, 1996).

Nickel

Ni^{2+} (10 mg/L) was accumulated in Indian mustard roots with a bioaccumulation coefficient of 208 after 24 hours (Dushenkov *et al.*, 1995). Ni was accumulated by the roots of Indian mustard *(Brassica juncea)* in water concentrations of about 20 to 2,000 g/L, with bioaccumulation coefficients of 500 to 2,000 (Salt *et al.*, 1997). Ni at concentrations of 1 to 16 mg/L was accumulated by water milfoil *(Myriophyllum spicatum)* with a minimum residual concentration of approximately 0.01 mg/L (Wang *et al.*, 1996).

Zinc

Zn^{2+} (100 mg/L) was accumulated in Indian mustard roots with a bioaccumulation coefficient of 131 after 24 hours (Dushenkov *et al.*, 1995). Zn at concentrations of 1 to 16 mg/L was accumulated by water milfoil *(Myriophyllum spicatum)* with a minimum residual concentration of approximately 0.1 mg/L (Wang *et al.*, 1996).

Chromium

Cr^{6+} (4 mg/L) was accumulated in Indian mustard roots with a bioaccumulation coefficient of 179 after 24 hours. The roots contained Cr^{3+}, indicating reduction of Cr^{6+} (Dushenkov *et al.*, 1995). Cr was accumulated by the roots of Indian mustard *(Brassica juncea)* in water concentrations of about 20 to 2000 g/L, with bioaccumulation coefficients of 100 to 250 (Salt *et al.*, 1997).

Radionuclides

Uranium

Uranium accumulation was studied using sunflowers in bench-scale and pilot-scale engineered systems (Dushenkov *et al.*, 1997).

Cesium

Cesium was used with sunflowers in bench-scale and pilot-scale engineered systems (Dushenkov *et al.*, 1997). Cs was accumulated in the roots of Indian mustard *(Brassica juncea)* in water concentrations of approximately 20 to 2,000 g/L, with bioaccumulation coefficients of 100 to 250 (Salt *et al.* 1997).

Strontium (Sr)

Sunflowers were used for the accumulation of strontium (Dushenkov *et al.*, 1997). Initially the bioaccumulation coefficient was 200 g/L, which decreased to 35 g/L within 48 hours, then down to 1 g/L by 96 hours. Sr was accumulated in the roots of Indian mustard *(Brassica juncea)* in water concentrations of approximately 20 to 2,000 g/L (Salt *et al.*, 1997).

Rhizofiltration occurs within the root zone in water. For rhizofiltration to occur, the water must come into contact with the roots. Engineered systems can be designed to maximise this contact zone by matching the depth of the unit to the depth of the roots. Groundwater may be extracted from any depth and piped to an engineered hydroponic system for exsitu treatment. The depth of treatable groundwater is a function of the extraction system, not the rhizofiltration treatment system.

For *in-situ* technologies, such as natural water bodies, the depth of the roots might not be the same as the depth of the water body. The water must be adequately circulated in such cases to ensure complete treatment, which is likely to become more difficult as the depth of the water increases.

Genetic Engineering in Rhizoremtionedia

Researchers at the Sainsbury Laboratory (Norwich, UK), in collaboration with the Institute of Genetics and Cytology (Minsk, Belarus), have successfully produced rhamnolipids in transgenic plants expressing genes derived from the soil bacterium *Pseudomonas aeruginosa*.

Table: Plant Species Applied for Rhizofilteration

Plant Species	Contaminants	References
Sunflowers *(Helianthus annuus)*	Cr^{6+}, Mn, Cd, Ni, Cu, U, Pb, Zn and Sr	Salt *et al.*, 1995
Brassica juncea and Bean *(Phaseolus coccineus)*	U and Ce	Dushenkov *et al.*, 1997
Aquatic plants		
Water hyacinth *(Eichhomia crassipes)*		Wang *et al.*, 1996,
Pennyworth (*Hydrocotyle umbellata*).	Ni, Cr, Cd, Cu, Zn and Pb	Dushenkov *et al.*, 1995
Duckweed (*Lemna minor*), Water velvet *(Azolla pinnata)* Water milfoil *(Myriophyllum spicatum)*		

Transgenic *Artzbidopsis* and tobacco plants were created expressing individual or combinations of different *P. aeruginosa* genes (e.g. *rhl*A and *rhl*B) responsible for the biosynthesis of rhamnolipids. All such plants had normal phenotypes, and showed successful production of rhamnolipids, confirmed by using TLC and MS analysis of plant extracts in comparison to commercially available bacterially derived rhamnolipid preparations. Transgenic plants producing rhamnolipids were found to exhibit dramatically enhanced growth and contamination tolerance on crude oil and heavy-metal contaminated soils, and they significantly enhanced the biodegradation of crude oil.

They flourished in soil spiked with 2.4 per cent number 4 fuel oil, breaking down nearly 25-35 per cent of oil in 45 days. Similar non-engineered plants were stunted and they accumulated oil. Increased resistance/tolerance was also exhibited to bacterial and fungal pathogens (including *Phytophthora infestans)* (Brychkova *et al.*, 2004).

Why Use Transgenic Plants in Phytoremediation?

To Improve Plant Performance: Initially, phytoremediation trials were performed using plants assumed or known to accumulate metals and/or to possess metal tolerance, e.g. Silene vulgaris (Moench), Brassicaceae crop plants like *Brassica oleracea* and *Raphanus sativus* and metal hyperaccumulators like *Thlaspi caerulescens* and *Alyssum* spp (Brown *et al.*, 1994). Metal hyperaccumulators were most efficient at metal removal in these field trials. Based on the results from pilot field trials, it was concluded that, for phytoextraction to become a viable technology, dramatic improvements would be required in either hyperaccumulator biomass yield or non-accumulator metal accumulation (McGrath *et al.*, 2000). These observations underline one of the potential benefits of transgenic approaches in phytoremediation.

Exploit Genetic Biodiversity Outside the Plant Kingdom

Naturally occurring hyperaccumulation and tolerance are largely specific for one trace element polluted soils often contain high levels of several contaminant trace elements, so that phytoremediation is likely to require the use of plants engineered to at least tolerate and, ideally, to accumulate several trace elements (Chaney *et al.*, 1997). Soils polluted with arsenic, cadmium, lead or mercury provide or major targets for remediation. To date, no plants have been identified which reproducibly hyperaccumulate lead or mercury. In most naturally occurring tolerant plants studied to date, tolerance to arsenic or lead appears to be based on exclusion from the plant. The development of a phytoremediation technology for some trace elements is thus likely to require the transfer of genes into plants across species borders.

Although we know little about the molecular basis of trace element detoxification and hyperaccumulation in plants, a number of trace element detoxification systems from bacteria and yeast have been characterised genetically and functionally at the molecular level. Despite the difficulty in predicting the effects of microbial genes in a complex multicellular organism like a plant, the successful introduction of a modified bacterial mercuric ion-reductase gene into yellow poplar *(Liriodendron tulipifera)* and *Nicotiana tabacum* demonstrates that bacterial genes may be extremely valuable in phytoremediation (Nies, 1999).

With the aid of genetic engineering, plants can be genetically modified to substantially improve phytoremediation. Expression of several plant and bacterial genes in transgenic plants has significantly enhanced these plant remediation systems. Several studies have successfully integrated bacterial genes into nuclear genomes to produce plants that were specifically engineered for phytoremediation of metal-polluted environments (Nies, 1999).

With respect to Hg, plants have been engineered with modified bacterial mercuric ion reductase *(merA)* and organomercurial lyase *(merB)* genes; these enzymes are capable of converting highly toxic methyl-Hg into the much less toxic Hg(0), which may then be volatilised. All of the attempts to genetically engineer plants with improved phytoremediation are mainly based on transformation of the nuclear genome. Ruiz *et al* (2003) engineered the chloroplast genomes of higher plants. Genes for mercuric ion reductase *(merA)* and organomercurial lyase *(merB)* were transferred from bacteria to chloroplast genes of higher plants.

Table: Genes introduced into plants and the effects of their expression on trace element tolerance, accumulation or volatilisation (Kramer and Chardonnens, 2001)

Gene	Product	Source	Target	Maximum observed effect	Root medium
merA	Hg(II) reductase	Gram-negative bacteria	*Liriodendron tulipifera*	T50μMHgCl$_2$; 500 mg HgCl$_2$ kg'[1]	Agar, soil
			Nicotiana tabacum	V Hg-volatilisation rate increased ten-fold	Agar
merA *merB*	Hg(II) reductase, Organomercurial lyase	Gram-negative bacteria	*Arabidopsis thaliana*	T10μMCH$_3$HgCl (>40-fold)	Agar
				V Up to 59 pg Hg(O) mg^{-1} fresh biomass min"	Solution
*APS*1	ATP sulfurylase	*A. thaliana*	*Brassica juncea*	A Two-fold increase in Se concentrations	Hydroponic
MT-I	MT	mouse	*N. tabacum*	T200μMCdCl$_2$ (20-fold)	Agar
*CUP*1	MT	*Saccharomyces cerevisiae*	*B. oleracea*	T400μMCdCl$_2$ (approx. 16-fold)	Hydroponic

Contd...

Gene	*Product*	*Source*	*Target*	*Maximum observed effect*	*Root medium*
ghs2	GSH synthase	*Escherichiacoli*	*B. juncea*	ACd concentrations 125%	Hydroponic
ghs1	Glu-Cys synthase	E *coli*	*B. juncea*	ACd concentrations 190%	Hydroponic
MCBP4	Cation channel	*M. tabacum*	*N. tabacum*	T250pMNiCl_2 (2.5-fold), Pb-sensitive APb concentrations 200%	Hydroponic
ZAT1	Zn transporter	*A. thaliana*	A *thaliana*	T Slight increase	Hydroponic

- *A:* Accumulation in the shoot
- *GSH:* glutathione
- MT: Metallothioneins
- *T:* tolerance
- *V*: volatilisation

Transformation of chloroplast genes offer several advantages over nuclear transformation, i.e. very high levels of transgene expression (up to 46 per cent (w/w) of total protein; uniparental plastid gene inheritance (in most crop plants) that prevents pollen transmission of foreign DNA; the absence of gene silencing and positioning effect and the ability to express multiple genes in a single transformation event.

There are number of examples illustrating the transfer of pollutant degradaing genes from bacteria to plant. Cotton plants were transformed with a bacterial gene degrading herbicide 2,4-dichlorophenoxyacetic acid (Lyon, 1993). The gene for 2,4-D monooxygenase, a bacterial enzyme that degrades the broadleaf herbicide 2,4-dichlorophenoxyacetic acid (2,4-D), was introduced into cotton plants by *Agrobacterium-mediated* transformation.

Several varieties of transgenic poplar containing cytochrome P-450 2E1 have been constructed and are undergoing tests. Strategies for improving public acceptance and safety of transgenic poplar for chlorinated hydrocarbon phytoremediation are being developed. Several varieties of transgenic poplar have been developed containing the human cytochrome 2E1 gene under the control of different promoters and some plants with NADPH P450 oxidoreductase and cytochrome b_5 in addition to 2El(Banerjee *et al.,* 2002).

Transgenic tobacco plants expressing pentaerythritol tetranitrate reductase, an enzyme derived from an explosive-degrading bacterium that enables degradation of nitrate ester and nitroaromatic explosives, were generated by French *et al.,* 1999. Seeds from transgenic plants were able to germinate and grow in the presence of 1 mM glycerol trinitrate (GTN) or 0.05 mM trinitrotoluene, at concentrations that inhibited germination and growth of wild-type seeds. Transgenic seedlings grown in liquid medium with 1 mM GTN showed more rapid and complete denitration of GTN than wild-type seedlings. This example suggests that transgenic plants expressing microbial degradative genes may provide a generally applicable strategy for bioremediation of organic pollutants in soil.

Future Prospects

Appealing features of rhizodegradation include destruction of the contaminant *in-situ,* complete mineralisation of organic contaminants and translocation of the compound to the plant or atmosphere is less likely than with other phytoremediation technologies since degradation occurs at the source of the contamination. Harvesting of the vegetation is not necessary since there is contaminant degradation within the soil, rather than contaminant accumulation within the plant. Root penetration throughout the soil may allow a significant percentage of the soil to be contacted. However, at a given time only a small percentage of the total soil volume is in contact with living roots.

It can take a long time for root dieback and root growth into new areas of the soil for contact with most of the soil to occur. Also, inhospitable soil conditions or areas of high contaminant concentrations can decrease root penetration, leading to some portions of the soil never being contacted by roots.

Plants offer many advantages over bacteria as agents for bioremediation; however, they typically lack the degradative capabilities of specially selected bacterial strains. Transgenic plants expressing microbial degradative enzymes could combine the advantages of both systems.

The potential for using genetic resources in transgenic phytoremediation approaches has not yet been fully explored. Furthermore, a number of transgenic plants have been engineered to contain large amounts of recombinant proteins with a possible role in chelation, assimilation or membrane transport of trace elements. Transgenic plants carrying genes responsible for degradation of various recalcitrant compounds have also been developed. Virtually no data are yet available on the performance of these transgenic plants on soil substrates or under field conditions, where trace element bioavailability is substantially lower. It is often difficult to predict the effects of the expression of a transgene at the level of the whole plant. An improved understanding of metal homoeostasis in plants will be vital for the development of successful phytoremediation technologies. Future strategies are likely to involve the introduction of several transgenes in tandem, the use of tissue-specific and inducible promoters, the manipulation of regulatory processes in plant metal homoeostasis and the generation of hybrids between metal hyperaccumulators and related high-biomass crop plants.

7

Biodegradation

Aerobic Granular Reactor

Aerobic granular reactor is a type of waste treatment facility. Waste Water Treatment Plants (WWTP) based on activated sludge often cover large surface areas, mainly caused by the large settling tanks. To be able to build compact WWTP's biomass can be grown as biofilms on a carrier material or as fast settling aerobic granular sludge without a carrier. Recent research showed the advantages of a discontinuously fed system, in which it is possible to grow stable granulated sludge under aerobic circumstances. Simultaneous COD, N and P removal can be easily integrated in a discontinuous fed system. Because of the high settling capacity of the granules, the use of traditional settler is unnecessary and therefore, the installation can be built very compact, needing only 20 per cent of the surface area of conventional activated sludge systems.

Bioaccumulation

Bioaccumulation occurs when an organism absorbs a toxic substance at a rate greater than that at which the substance is lost. Thus, the longer the biological half-life of the substance the greater the risk of chronic poisoning, even if environmental levels of the toxin are very low.

This is one reason why chronic poisoning is a common aspect of environmental health in the workplace. As people spend so much time, for so many years in these environments, very low levels of toxins can be lethal over time.

An example of poisoning in the workplace can be seen from the phrase "as mad as a hatter". The process for stiffening the felt used in making hats involved mercury, which forms organic species such as methylmercury, which is lipid soluble, and tends to accumulate in the brain resulting in mercury poisoning.

Other lipid (fat) soluble poisons include tetra-ethyl lead compounds (the lead in leaded petrol), and DDT. These compounds are stored in the body's fat, and when the fatty tissues are used for energy, the compounds are released and cause acute poisoning.

Strontium 90, part of the fallout from atomic bombs, is mistaken by the human body for calcium, and is laid down in the bone, where its radiation can cause damage for a long time.

Naturally produced toxins can also bioaccumulate. The marine algal blooms known as "red tides" can result in local filter feeding organisms such as mussels and oysters becoming toxic; coral fish can be responsible for the poisoning known as ciguatera when they accumulate a toxin called ciguatoxin from reef algae.

Some animal species exhibit bioaccumulation as a mode of defence; by consuming toxic plants or animal prey, a species may accumulate the toxin which then presents a deterrent to a potential predator. One example is the tobacco hornworm, which concentrates nicotine to a toxic level in its body as it consumes tobacco plants.

Other compounds that are not normally considered toxic can be accumulated to toxic levels in organisms. The classic example is of Vitamin A, which becomes concentrated in carnivore livers of, e.g. polar bears: as a pure carnivore that feeds on other carnivores (seals), they accumulate extremely large amounts of Vitamin A in their livers. It was known by the native peoples of the arctic that the livers should not be eaten, but arctic explorers have suffered Hypervitaminosis A from eating the bear livers (and there has been at least one example of similar poisoning of Antarctic explorers eating husky dog livers).

Biodegradability Prediction

Biodegradability prediction is biologically-inspired computing and attempts to predict biodegradability of anthropogenic materials in the environment. Demand for biodegradability prediction is expected to increase with governments stepping up environmental regulations.

Example:

- Development of quantitative structure-activity relationship (QSARs) for biodegradation, for instance, biochemical oxygen demand for chemicals released into the environment with the aid of machine learning and other artificial intelligence methods
- The University of Minnesota Biocatalysis and Biodegradation Database (UM-BBD), which contains information on microbial biocatalytic reactions and biodegradation pathways for primarily xenobiotic, chemical compounds. One of its many features allows the prediction of microbial catabolic reactions using substructure searching, a rule-base, and atom-to-atom mapping.

Biodegradable Waste

Biodegradable waste is a type of waste, typically originating from plant or animal sources, which may be broken down by other living organisms.

It can be commonly found in municipal solid waste (sometimes called Biodegradable Municipal Waste [BMW]) as:

- Green waste
- Food waste
- Paper waste
- Biodegradable plastics

Other Biodegradable Waste Includes:

- Human waste
- Manure
- Sewage
- Slaughterhouse waste

Uses of Biodegradable Waste

Biodegradable waste is a little recognised resource. Through correct waste management, often using the two key processes of anaerobic digestion and composting, it can be converted into valuable products.

Anaerobic digestion converts biodegradable waste into several products, including biogas, which can be used to generate renewable energy or heat for local heating, and soil amendment (digestate). Composting converts biodegradable waste into compost.

Biodegradable Waste and Global Warming

Biodegradable waste is an important substance due to its links with global warming. When it is disposed of in landfills, it breaks down under uncontrolled anaerobic conditions. This produces landfill gas which, if not harnessed, escapes into the atmosphere. Landfill gas contains methane, a more potent greenhouse gas than carbon dioxide.

The European Union Landfill Directive puts key requirements on member states for the management of biodegradable waste in order to stop global warming.

Methanogenesis

Methanogenesis or biomethanation is the formation of methane by a group of microbes known as methanogens. Organisms capable of producing methane have been identified only from the kingdom Archaea, a group phylogenetically distinct from both eukaryotes and bacteria, although many methanogenic organisms live in close association with anaerobic bacteria. The production of methane is an important and widespread form of microbial metabolism. In most environments, it is the final step in the decomposition of biomass. Recently, it has been demonstrated that leaf tissues of living plants emit methane. Although the mechanism by which such methane production occurs is, as yet, unknown, the implications are far-reaching; this is an example of methanogenesis occurring in non-microbes, presumably under aerobic conditions. Since not much is known about the pathways employed in plants in producing methane, the rest of this article focuses on the well-studied production of methane by microbes.

Biochemistry of Methanogenesis

Methanogenesis is a form of anaerobic respiration. Methanogens do not use oxygen to breathe; in fact, oxygen inhibits the growth of methanogens. The terminal electron acceptor in methanogenesis is not oxygen, but carbon. The carbon can occur in a small number of organic compounds, all with low molecular weights. The two best described pathways involve the use of carbon dioxide and acetic acid as terminal electron acceptors:

$$CO_2 + 4\,H_2 \rightarrow CH_4 + 2H_2O$$

$$CH_3COOH \rightarrow CH_4 + CO_2$$

However, methanogenesis has been shown to use carbon from other small organic compounds, such as formic acid, methanol, methylamines, dimethyl sulphide, and methanethiol.

The biochemistry of methanogenesis is relatively complex, involving the following coenzymes and cofactors: F430, coenzyme B, coenzyme M, methanofuran, and methanopterin.

Occurrence of Methanogens

Organisms capable of methogensis are called methanogens. Microbes performing methanogenesis have no nucleus or membrane-bound organelles (i.e. they are prokaryotes). Methanogens are considered to be a very old group of organisms, being members of the archaebacteria, also known as archaea (depending on what taxonomic system is being used).

Methanogens cannot exist in the presence of oxygen, so they are only found in environments in which the oxygen has been depleted. Most commonly, these are environments experiencing the decay of organic matter, such as wetland soils, the digestive tracts of animals, and aquatic sediments. Methanogenesis also occurs in areas where oxygen and decaying organic matter are both absent, such as the terrestrial deep subsurface, deep-sea hydrothermal vents, and oil reservoirs.

Importance in Carbon Cycle

Methanogenesis is the final step in the decay of organic matter. During the decay process, electron acceptors (such as oxygen, ferric iron, sulphate, nitrate, and manganese) become depleted, while hydrogen (H_2) and carbon dioxide accumulate. Light organics produced by fermentation also accumulate. During advanced stages of organic decay, all electron acceptors become depleted except carbon dioxide. Carbon dioxide is a product of most catabolic processes, so it is not depleted like other potential electron acceptors.

Only methanogenesis and fermentation can occur in the absence of electron acceptors other than carbon. Fermentation only allows the breakdown of larger organic compounds, and produces small organic compounds. Methanogenesis effectively removes the semi-final products of decay: hydrogen, small organics, and carbon dioxide. Without methanogenesis, a great deal of carbon (in the form of fermentation products) would accumulate in anaerobic environments.

Methanogenesis is useful to humanity. Through methanogenesis, organic waste can be converted to useful methane "biogas." Methanogenesis occurs in the guts of humans and other animals. While methanogenesis is not believed to be necessary for human digestion, it is required for the nutrition of ruminant animals, such as cattle and goats. In the rumen (known incorrectly as the "second stomach" possessed by some animals), anaerobic organisms (including methanogens) digest cellulose into forms usable by the animal. Without the microbes of the rumen, cattle cannot survive without being fed a special diet.

Reverse Methanogenesis

Methanogens can also utilise methane as a substrate in conjunction with the reduction of sulphate and nitrate.

Thermophile

A thermophile is an organism – a type of extremophile – which thrives at relatively high temperatures, above 45°C. Many thermophiles are archaea.

Thermophiles have been found in various geothermally heated regions of the Earth such as hot springs like those in Yellowstone National Park and deep sea hydrothermal vents, as well as decaying plant matter such as peat bogs and compost.

As a prerequisite for their survival, thermophiles contain enzymes that can function at high temperature. Some of these enzymes are used in molecular biology (for example heat-stable DNA polymerases for polymerase chain reaction), and in washing agents.

Thermophiles are classified into obligate and facultative thermophiles: obligate thermophiles (also called *extreme thermophiles*) require such high temperatures for growth, while facultative thermophiles (also called *moderate thermophiles*) can thrive at high temperatures but also at lower temperatures (below 50°C). Hyperthermophiles are particular extreme thermophiles of which optimal temperatures are above 80°C.

Importance of Enzymes from Thermophiles

The enzymes isolated from some extremophiles have proven to be of great use in the biotechnology industry, able to function under conditions that would denature enzymes taken from most "normal" organisms. .

The most commonly used DNA polymerase for the polymerase chain reaction technique is Taq DNA polymerase, originally isolated from *Thermus aquaticus*, a bacterial species found in surface aquatic locations such as Yellowstone National Park hot springs.

Thermococcus litoralis was isolated from a deep sea hydrothermal vent. This DNA polymerase is marketed as "Vent" polymerase. Another heat stable polymerase comes from the organism *Pyrococcus furiosus*, (Pfu). This organism grows optimally at temperatures between 105-113°C, making it a hyperthermophile.

Taq DNA polymerase is adequate for most PCR, but one study (Hamilton *et al.*, 2001). reported that higher fidelity thermostable DNA polymerases such as Vent account for as much as 30 per cent of DNA polymerase sales.

In addition, the study of proteins from thermophilic organisms has provided important insight into the mechanism of protein folding because these proteins must be stable at temperatures that would denature typical proteins. Therefore, understanding how thermophilic proteins have evolved to be stable can yield information about the functional modulation of folding landscapes.

Some Thermophilic Organisms

- *Pyrococcus furiosus*
- *Thermus aquaticus*
- *Thermus thermophilus*
- *Chloroflexus aurantiacus* (photosynthetic bacterium)
- *Thermococcus litoralis*

- *Pyrodictium abyssi* (archaea)
- *Bacillus stearothermophilus*
- *Thermobifida fusca* (bacterium often found in decaying plant matter)

The Pompeii worm survives the scalding temperatures surrounding deep-sea hydrothermal vents thanks to a symbiotic relationship with thermophilic bacteria.

Extreme Thermophiles

Thermophiles are heat-loving (hence the name), with an optimum growth temperature of 50°C or more, a maximum of up to 70°C or more, and a minimum of about 20°C, but these are only approximate. Some extreme thermophiles (hyperthermophiles) require a very high temperature (80°C to 105°C) for growth. Their membranes and proteins are unusually stable at these extremely high temperatures. Thus many important biotechnological process utilise thermophilic enzymes because of their ability to withstand intense heat.

Many of the hyperthermophiles Archea require elemental sulphur for growth. Some are anaerobes that use the sulphur as an electron acceptor during respiration instead of oxygen. Some are lithotrophs that oxidise sulphur to sulphuric acid as an energy source, thus requiring the microorganism to be adapted to very low pH (i.e. it is an acidophile as well as thermophile). These organisms are inhabitants of hot, sulphur-rich environments usually associated with volcanism, such as hot springs, geysers and fumaroles. In these places, especially in Yellowstone National Park, we find a zonation of microorganisms according to their temperature optima. Often these organisma are coloured, due to the presence of photosynthetic pigments.

Thermaer: The Second Generation ATAD

The First TRULY Aerobic Biosolids Reduction System: Results improved significantly with the introduction of Autothermal Thermophilic Aerobic Digestion (ATAD) several years ago. ATAD takes advantage of highly efficient thermophilic organisms naturally present in wastewater, optimising the environment for them to proliferate and dominate. This increases the temperature of the sludge as the thermophiles feed on other microorganisms. At these higher temperatures, the cell walls of the activated sludge rupture, releasing the now-dead mesophilic contents and providing a feast for the thermophiles. The metabolism of the thermophiles is extremely high, yet the net yield is low, resulting in a significant reduction of volatile solids to produce a pathogen-free end product. On the downside, due to their inherent inflexibility, traditional ATAD processes often produce excess foam and unacceptable odours.

The first challenge the Thermal Process team undertook was to develop a more reliable, more efficient process for the production of high quality biosolids. The result is the proprietary ThermAer process. ThermAer is second-generation Autothermal Thermophilic Aerobic Digestion process that improves on existing technology in several ways. It gives you the best of all worlds — a *high quality* pathogen-free product — *plus* controlled foam, the elimination of foul odours through prevention or treatment as appropriate, and superior reduction of volatile solids. ThermAer is the only Autothermal Thermophilic *Truly* Aerobic Digestion process.

8

Microbial Transformation

Microbial Growth Theory

The dominant microbes responsible for the oxidation of biodegradable carbonaceous pollutants in biotreaters are chemoheterotrophic bacteria. Such bacteria reproduce by binary fission. When the necessary requirements for bacterial growth are satisfied, the growth rate of a bacterial culture can be expressed as:

$$dx/dt = \mu x$$

where x = microbial cell mass

μ = specific growth rate coefficient

This equation is acceptable for describing binary fission; the rate of increase is proportional to the mass of organisms. However, this equation lets the mass increase without limit. This must be false because no living things can grow after their supply of food is exhausted.

We can advance our understanding of microbial growth by making the specific growth rate a function of the concentration of some nutrient.

Growth-limiting Nutrient

Living cells require several types of nutrients. These may supply carbon-energy, building blocks for biochemicals, or trace factors such as vitamins or hormones. All cells require relatively large amounts of carbon, nitrogen, sulphur, phosphorus, oxygen, hydrogen, etc. Except in very rare cases in which the recipe for the medium has been carefully adjusted, one nutrient will be exhausted before the others; this is termed the "growth-limiting nutrient".

It is not simply mass that determines the growth-limiting nutrient, but rather the proportion. Something present at low concentrations can be growth-limiting if it is the first to be exhausted. Of several equations that relate specific growth rate coefficient to concentration of limiting nutrient, the Monod equation is most popular and usually fits actual data quite well:

$$\mu = \mu \frac{S}{K_s + S}$$

where μ = Specific growth rate coeff.

μ = Maximum growth rate coeff.

s = Concentration of limiting nutrient

K_S = Half-saturation coeff.

Chemotroph

Chemotrophs are organisms that obtain energy by the oxidation of electron donating molecules in their environments. These molecules can be organic (organotrophs) or inorganic (lithotrophs). The chemotroph designation is in contrast to phototrophs which utilise solar energy. Chemotrophs can be either autotrophic or heterotrophic.

- Chemoautotrophs (or chemotrophic autotroph), in addition to deriving energy from chemical reactions, synthesise all necessary organic compounds from carbon dioxide. Chemoautotrophs generally only use inorganic energy sources. Most are bacteria or archaea that live in hostile environments such as deep sea vents and are the primary producers in such ecosystems. Evolutionary scientists believe that the first organisms to inhabit Earth were chemoautotrophs that produced oxygen as a by-product and later evolved into both aerobic, animal-like organisms and photosynthetic, plant-like organisms. Chemoautotrophs generally fall into several groups: methanogens, halophiles, sulphur reducers, nitrifiers, anammoxbacteria and thermoacidophiles.

- Chemoheterotrophs (or chemotrophic heterotrophs) must ingest organic building blocks that they are incapable of creating on their own. Most chemoheterotrophs derive energy from organic molecules like glucose.

The names of these primary nutritional groups are built from Greek roots. "Chemo" means "chemical" and "troph" means "nourishment". "Auto" means "self" and "hetero" means "other".

Endophyte

An endophyte is an organism that lives within a plant for at least part of its life without causing apparent disease. Endophytes are ubiquitous and have been found in nearly all plants studied to date.

However, most of these endophyte/plant relationships are not well understood; the type of symbiosis exhibited between plant and endophyte is often related to the reproductive mode of the endophyte. For example, vertically transmitted endophytes (i.e. transmited from parent to its offspring directly) are asexual and transmit via fungal hyphae penetrating the host's seeds (e.g. *Neotyphodium*). Evolutionary theory predicts these endophytes to evolve towards mutualism, since their reproductive fitness is intimately tied to that of their host plant. Conversely, horizontally transmitted endophytes are sexual and transmit via spores that can be spread by wind and/or insect vectors. Therefore, these endophytes can evolve a more pathogenic lifestyle since they can escape their host plant.

Exogenous Bacteria

Exogenous bacteria are bacteria introduced to closed biological systems from the external world. They exist in water, earth, and the air. Examples are cholera, *Legionella,* salmonella, Lyme disease, tuberculosis, and bactillus anthracis. *Endogenous* bacteria are part of our normal internal flora.

Extremophile

An extremophile is an organism, usually unicellular, which thrives in or requires 'extreme' conditions that would exceed optimal conditions for growth and reproduction in the majority of mesophilic terrestrial organisms.

Most extremophiles are microbes. The domain Archaea is known for widespread extremophilia, but extremophiles are present in numerous and diverse genetic lineages of both the bacteria and archaea. Although the terms archaea and extremophile are occasionally used interchangeably, there are many mesophile archaeans and many extremophile bacteria. Additionally, not all extremophiles are unicellular. Examples of extremophilic metazoa are the Pompeii worm, the psychrophilic Grylloblattodea (insects), antarctic krill (crustaceans) and the Tardigrade.

Types of Extremophiles

There are many different classes of extremophiles, each corresponding to the way its environmental niche differs from those of the majority of terrestrial mesophile organisms. These classifications are not exclusive. Many extremophiles fall under multiple categories. For example, organisms living inside hot rocks deep under Earth's surface are both *thermophilic* and *barophilic.*

- Acidophile: An organism with an optimum pH level at or below pH 3.

- Alkaliphile: An organism with optimal growth at pH levels of 9 or above.
- Barophile: Bacteria which live in environments characterised by high gas or liquid pressure; synonymous with piezophile.
- Endolith: An organism that lives in microscopic spaces within rocks, such as pores between aggregate grains. These may also be called cryptoendoliths. This term also includes organisms populating fissures, aquifers, and faults filled with groundwater in the deep subsurface.
- Halophile: An organism requiring at least 2M of salt, NaCl, for growth.
- Hyperthermophile: An organism that can thrive at temperatures between 80-121°C, such as those found in hydrothermal systems.
- Hypolith: An organism that lives inside rocks in cold deserts.
- Lithoautotroph: An organism (usually bacteria) whose sole source of carbon is carbon dioxide and exergonic inorganic oxidation (chemolithotrophs) such as *Nitrosomonas europaea*. These organisms are capable of deriving energy from reduced mineral compounds like pyrites, and are active in geochemical cycling and the weathering of parent bedrock to form soil.
- Metalotolerant: capable of tolerating high levels of dissolved heavy metals in solution, such as copper, cadmium, arsenic, and zinc.
- Oligotroph: An organism capable of growth in nutritionally limited environments.
- Osmophile: An organism capable of growth in environments with a high sugar concentration.
- Piezophile: An organism that lives optimally at high hydrostatic pressure. Common in the deep terrestrial subsurface, as well as in oceanic trenches.

- Polyextremophile: An organism that qualifies as an extremophile under more than one category.
- Psychrophile/Cryophile: An organism that grows better at temperatures of 15°C or lower. Common in cold soils, permafrost, polar ice, cold ocean water, and in/under alpine snowpack.
- Radioresistant: resistant to high levels of ionising radiation, most commonly ultraviolet radiation but also includes organisms capable of resisting nuclear radiation.
- Thermophile: An organism that can thrive at temperatures between 60-80°C.
- Xerophile: An organism that can grow in extremely dry, desiccating conditions. This type is exemplified by the soil microbes of the Atacama Desert.

Extremophiles and Astrobiology

Astrobiology is the field concerned with forming theories about the distribution, nature, and future of life in the universe. In it, microbial ecologists, astronomers, planetary scientists, geochemists, philosophers, and explorers cooperate to constructively guide the search for life on other planets. Astrobiologists are particularly interested in studying extremophiles, as many organisms of this type are capable of surviving in environments similar to those known to exist on other planets. For example, Mars may have regions in its deep subsurface permafrost that could harbour endolith communities. The subsurface water ocean of Jupiter's moon Europa may harbour life, especially at hypothesised hydrothermal vents at the ocean floor.

Geophilic

Geophillic means soil loving or preferring the soil. Geophilic organisms are usually recovered from the soil but occasionally infect humans and animals. They cause a marked inflammatory reaction, which limits the spread of the infection and may lead to a spontaneous cure but may also leave scars.

Lithoautotrophy

Lithoautotrophy is a special type of chemoautotrophy found in Archaea and Bacteria. Lithoautotrophic organisms utilise inorganic compounds as energy sources. The word *lithoautotrophy* means "to feed it- or oneself from stone". In ancient Greek, *lithos* means stone or rock, *autos* self, and *trophein* to feed.

Mesophile

A mesophile is an organism that grows best in moderate temperature, neither too hot nor too cold, typically between 25 and 40°C (77 and 104°F). The term is mainly applied to microorganisms.

The habitats of these organisms include soil, the human body, animals, etc. The optimal temperature of many pathogenic mesophiles is 37°C (98°F), the normal human body temperature. Mesophilic organisms have important uses in food preparation especially in cheese and yoghurt making and in beer and wine making.

Organisms that prefer cold environments are termed psychrophilic, those preferring warmer temperatures are termed thermophilic and those thriving in extremely hot environments are hyperthermophilic.

Methylotroph

Methylotrophs are a diverse group of microorganisms that can utilise reduced one-carbon compounds, such as methanol or methane, as the carbon source for their growth. Some methylotrophs can degrade the greenhouse gas methane. The methylotroph *Methylococcus capsulatus* is used to reduce methane and other pollutants. The abundance, purity, and low price of methanol compared to commonly used sugars make methylotrophs competent organisms for production of amino acids, vitamins, recombinant proteins, single-cell proteins, co-enzymes and cytochromes.

Examples of Methylotrophs

- Genus *Methanosarcina*
- *Methylococcus capsulatus*
- *Pichia pastoris*

Microaerophile

Microaerophilic organisms are a specific type of organism that requires oxygen to survive, but requires environments containing lower levels of oxygen than are present in the atmosphere (~20 per cent concentration).

Examples include *Borrelia burgdorferi*, a species of spirochaete bacteria that causes Lyme disease in humans, and *Helicobacter pylori*, a species of proteobacteria that has been linked to peptic ulcers and some types of gastritis.

Osmotrophy

Osmotrophy is the uptake of dissolved organic compounds by osmosis for nutrition. Organisms that use osmotrophy are osmotrophs. Some mixotrophic microorganisms use osmotrophy to derive energy.

Psychrophile

Psychrophiles or Cryophiles (adj. cryophilic) are extremophilic organisms that are capable of growth and reproduction in cold temperatures. They can be contrasted with thermophiles, which thrive at unusually hot temperatures. The environments they inhabit are ubiquitous on Earth, as a large fraction of our planetary surface experiences temperatures lower than 15°C. They are present in alpine and arctic soils, high-latitude and deep ocean waters, arctic ice, glaciers, and snowfields. They are of particular interest to astrobiology, the field dedicated to the formulation of theory about the possibility of extraterrestrial life, and to geomicrobiology, the study of microbes active in geochemical processes.

Psychrophiles utilise a wide variety of metabolic pathways, including photosynthesis, chemoautotrophy (also sometimes known as lithotrophy), and heterotrophy, and form robust, diverse communities. Most psychrophiles are bacteria or archaea, and psychrophily is present in widely diverse microbial lineages within those broad groups. Additionally, recent research has discovered novel groups of psychrophilic fungi living in oxygen-poor areas under alpine snowfields. A further group of eukaryotic cold-adapted organisms are snow algae, which can cause watermelon snow. Psychrophiles are characterised by lipid cell membranes chemically resistant to the stiffening caused by extreme cold, and often create protein 'antifreezes' to keep their internal space liquid and protect their DNA even in temperatures below water's freezing point.

Types of Psychrophiles

There are generally considered to be two groups of psychrophiles: *"obligate"* psychrophiles, and a second group, the *"facultative psychrophiles"*, that are sometimes referred to as *psychrotrophs* by food microbiologists.

Obligate Psychrophiles

Obligate psychrophiles are those organisms having a growth temperature optimum of 15°C or lower and cannot grow in a climate beyond a maximum temperature of 20°C. They are largely found in icy places (such as in Antarctica) or at the freezing bottom of the ocean floor. This separation is becoming more difficult as more organisms with a fairly large growth temperature range are discovered.

Facultative Psychrophiles

Facultative psychrophiles can grow at 0°C up through approximately 40°C, and exist in much larger numbers than obligate psychrophiles. They are generally not able to grow much below 0°C, though they may maintain basic functioning. They have evolved to tolerate cold, but they are not as physiologically specialised as obligate psychrophiles and are usually not found in the very coldest of environments. They are of particular significance to food microbiologists as they can grow in refrigerated environments and thereby cause food spoilage.

Examples

The following are species with psychrophilic natures

- The genus *Moritella* is mostly made up of psychrophiles.
- *Leifsonia aurea* found in the Antarctic.
- *Methanococcoides burtonii* was isolated from Ace Lake, Antartica.

Saprotroph

A saprotroph (or saprobe) is an organism that obtains its nutrients from non-living organic matter, usually dead and decaying plant or animal matter, by absorbing soluble organic compounds. Since saprotrophs cannot make food for themselves, they are considered a type of heterotroph. They include many fungi (the rest being parasitic, commensal or mutualistic symbionts), bacteria, and protozoa. Animal scavengers, such as dung beetles, vultures, and a few unusual non-photosynthetic plants are also sometimes referred to as saprotrophs, but are more commonly called saprophages.

Saprophyte is an older term that is now considered obsolete. The suffix -phyte means "plant". However, there are no truly saprotrophic organisms that are embryophytes, and fungi and bacteria are no longer placed in the Plant Kingdom. Plants that were once considered saprophytes, such as non-photosynthetic orchids and monotropes, are now known to be parasites on other plants. They are termed myco-heterotrophs because a mycorrhizal fungus connects the parasitic plant with its host plant.

Some saprotrophic organisms are useful scavengers, and in sewage farms and refuse dumps break down organic matter into nutrients easily assimilable by green plants.

Factors that Influence Microbial Growth

Many factors must be evaluated for each specific food

when making decisions on whether it needs time/temperature control for safety. These can be divided into intrinsic and extrinsic factors. Intrinsic factors are those that are characteristic of the food itself; extrinsic factors are those that refer to the environment surrounding the food. The need for time/temperature control is primarily determined by: (1) the potential for contamination with pathogenic microorganisms of concern (including processing influences), and (2) the potential for subsequent growth and/or toxin production.

Most authorities are likely to divide foods among three categories based on an evaluation of the factors described below: those that do not need time/temperature control for protection of consumer safety; those that need time/temperature control; and those where the exact status is questionable. In the case of questionable products, further scientific evidence—such as modelling of microbial growth or death, or actual microbiological challenge studies—may help to inform the decision.

Moisture Content

Microorganisms need water in an available form to grow in food products. The control of the moisture content in foods is one of the oldest exploited preservation strategies. Food microbiologists generally describe the water requirements of microorganisms in terms of the water activity (a_w) of the food or environment. Water activity is defined as the ratio of water vapour pressure of the food substrate to the vapour pressure of pure water at the same temperature:

$$a_w = p/po,$$

where p = vapour pressure of the solution and p_o = vapour pressure of the solvent (usually water). The a_w of pure water is 1.00 and the a_w of a completely dehydrated food is 0.00. The a_w of a food on this scale from 0.00-1.00 is related to the equilibrium relative humidity above the food on a scale of 0-100 per cent. Thus, per cent Equilibrium Relative Humidity (ERH) = $a_w \times 100$. The a_w of a food describes the degree to which water is "bound" in the food, its availability to participate in chemical/biochemical reactions, and its availability to facilitate growth of microorganisms.

Most fresh foods, such as fresh meat, vegetables, and fruits, have a_w values that are close to the optimum growth level of most microorganisms (0.97-0.99). The a_w can be manipulated in foods by a number of means, including addition of solutes such as salt or sugar, physical removal of water through drying or baking, or binding of water to various macromolecular components in the food. These food components will decrease a_w in the following order: ionic compounds > sugars, polyhydric alcohols, amino acids and other low-molecular-weight compounds > high-molecular-weight compounds such as cellulose, protein or starch.

Microorganisms respond differently to a_w depending on a number of factors. Microbial growth, and, in some cases, the production of microbial metabolites, may be particularly sensitive to alterations in a_w. Microorganisms generally have optimum and minimum levels of a_w for growth depending on other growth factors in their environments. One indicator of microbial response is their taxonomic classification. For example, Gram (-) bacteria are generally more sensitive to low a_w than Gram (+) bacteria. It should be noted that many bacterial pathogens are controlled at water activities well above 0.86 and only *S. aureus* can grow and produce toxin below a_w 0.90. It must be emphasised that these are approximate values because solutes can vary in their ability to inhibit microorganisms at the same aw value. To illustrate, the lower a_w limit for the growth of *Clostridium botulinum* type A has been found to be 0.94 with NaCl as the solute versus 0.92 with glycerol as the solute.

When formulating foods using a_w as the primary control mechanism for pathogens, it is useful to employ microbiological challenge testing to verify the effectiveness of the reduced aw when target aw is near the growth limit for the organism of concern.

Because a_w limits vary with different solutes or humectants, other measures may provide more precise moisture monitoring for certain products. For example, factors other than a_w are known to control the anti-botulinal properties of pasteurised processed cheese spreads (Tanaka and others 1986). Also, a_w may be used in combination with other factors to control pathogens in certain food products.

Care should be taken when analysing multi-components foods, because effective measurements of a_w may not reflect the actual value in a microenvironment or in the interface among the different components. In these cases, the a_w should be measured at the interface areas of the food, as well as in any potential microenvironment.

Table: Approximate a_w values of selected food categories.

Animal Products	a_w
fresh meat, poultry, fish	0.99 – 1.00
natural cheeses	0.95 – 1.00
pudding	0.97 - 0.99
eggs	0.97
cured meat	0.87 – 0.95
sweetened condensed milk	0.83
Parmesan cheese	0.68 – 0.76
honey	0.75
dried whole egg	0.40
dried whole milk	0.20
Plant Products	a_w
fresh fruits, vegetables	0.97 – 1.00
bread	~0.96
white	0.94 – 0.97
crust	0.30
baked cake	0.90 – 0.94
maple syrup	0.85
jam	0.75 – 0.80
jellies	0.82 – 0.94
uncooked rice	0.80 – 0.87
fruit juice concentrates	0.79 – 0.84

Contd...

Animal Products	a_w
fruit cake	0.73 – 0.83
cake icing	0.76 – 0.84
flour	0.67 – 0.87
dried fruit	0.55 – 0.80
cereal	0.10 – 0.20
sugar	0.19
crackers	0.10

Table: Approximate a_w values for growth of selected pathogens in food

Organism	*Minimum*	*Optimum*	*Maximum*
Campylobacter spp.	0.98	0.99	
Clostridium botulinum type E*	0.97		
Shigella spp.	0.97		
Yersinia enterocolitica	0.97		
Vibrio vulnificus	0.96	0.98	0.99
Enterohemorrhagic *Escherichia coli*	0.95	0.99	
Salmonella spp.	0.94	0.99	>0.99
Vibrio parahaemolyticus	0.94	0.98	0.99
Bacillus cereus	0.93		
Clostridium botulinum types A & B**	0.93		
Clostridium perfringens	0.943	0.95-0.96	0.97
Listeria monocytogenes	0.92		
Staphylococcus aureus growth	0.83	0.98	0.99
Staphylococcus aureus toxin	0.88	0.98	0.99

pH and Acidity

Increasing the acidity of foods, either through fermentation or the addition of weak acids, has been used as a preservation method since ancient times. In their natural state, most foods such as meat, fish, and vegetables are slightly acidic while most fruits are moderately acidic. A few foods such as egg white are alkaline. The pH is a function of the hydrogen ion concentration in the food:

$$pH = -\log_{10} [H^+]$$

Another useful term relevant to the pH of foods is the pKa. The term pKa describes the state of dissociation of an acid. At equilibrium, pKa is the pH at which the concentrations of dissociated and undissociated acid are equal. Strong acids have a very low pKa, meaning that they are almost entirely dissociated in solution. For example, the pH (at 25°C [77°F]) of a 0.1 M solution of HCl is 1.08 compared to the pH of 0.1 M solution of acetic acid, which is 2.6. This characteristic is extremely important when using acidity as a preservation method for foods. Organic acids are more effective as preservatives in the undissociated state. Lowering the pH of a food increases the effectiveness of an organic acid as a preservative. The type of organic acid employed can dramatically influence the microbiological keeping quality and safety of the food.

It is well known that groups of microorganisms have pH optimum, minimum, and maximum for growth in foods. As with other factors, pH usually interacts with other parameters in the food to inhibit growth. The pH can interact with factors such as aw, salt, temperature, redox potential, and preservatives to inhibit growth of pathogens and other organisms. The pH of the food also significantly impacts the lethality of heat treatment of the food. Less heat is needed to inactivate microbes as the pH is reduced (Mossel and others 1995).

Another important characteristic of a food to consider when using acidity as a control mechanism is its buffering capacity. The buffering capacity of a food is its ability to resist changes in pH. Foods with a low buffering capacity will change pH quickly in response to acidic or alkaline compounds produced by microorganisms as they grow. Meats, in general, are more buffered than vegetables by virtue of their various proteins. Titratable acidity (TA) is a better indicator of the microbiological stability of certain foods, such as salad dressings, than is pH. Titratable acidity is a measure of the quantity of standard alkali (usually 0.1 M NaOH) required to neutralise an acid solution. It measures the amount of hydrogen ions released from undissociated acid during titration. Titratable acidity is a particularly useful measure for highly buffered or highly acidic foods. Weak acids (such as organic acids) are usually undissociated and, therefore, do not directly contribute to pH. Titratable acidity yields a measure of the total acid concentration, while pH does not, for these types of foods.

In general, pathogens do not grow, or grow very slowly, at pH levels below 4.6; but there are exceptions. Many pathogens can survive in foods at pH levels below their growth minima. It has been reported that *C. botulinum* was able to produce toxin as low as pH 4.2, but these experiments were conducted with high inoculum levels (103-104 CFU/g up to 106 CFU/g), in soy peptone, and with the presence of *Bacillus* spp.

(Smelt and others 1982). The panel did not consider these results to be relevant to the foods under consideration in this report. It should also be noted that changes in pH can transform a food into one which can support growth of pathogens. For example, several botulism outbreaks have been traced to foods in which the pH increased due to mould growth. These are important considerations when determining the shelf life of a food formulation. Based on a comprehensive review of the literature, the panel concluded that a pH of 4.6 is appropriate to control spore-forming pathogens.

Among vegetative pathogens, *Salmonella* spp. are reported to grow at the lowest pH values; however, in a study by Chung and Goepfert (1970), the limiting pH was greatly influenced by the acidulant used. For example, when tryptone-yeast extract-glucose broth was inoculated with 104 CFU/ml of salmonellae, minimum pH values for growth ranged from 4.05 with hydrochloric and citric acids to 5.5 with propionic acid or acetic acid. Additionally, inoculum levels were unrealistically high (102-106 CFU/ml) for salmonellae in food systems. These investigators also noted that these results could not be extrapolated directly to food because the experiment was run in laboratory media under ideal temperature and a_w conditions and without the presence of competitive microorganisms.

Similarly, Ferreira and Lund (1987) reported that six out of 13 strains of *Salmonella* spp. representing 12 serovars could grow at pH 3.8 at 30°C (86°F) within 1-3 d, and at 20°C (68°F) in 3-5 d, when using HCl as an acidulant. Other reports note that certain acids at pH 4.5 inactivate salmonellae. The panel therefore concluded that using a Ph minimum of 4.0 for *Salmonella* spp. would not be scientifically substantiated for foods subject to Food Code requirements. Based on a comprenhensive review of the literature data, the panel also concluded that it would be scientifically valid to use a pH minimum of 4.2 to control for *Salmonella* spp. and other vegetative pathogens.

As with other intrinsic properties, when analysing multi-components foods, the pH should be measured not only for each component of the food but also for the interface areas among components and for any potential microenvironment.

Table: pH ranges of some common foods

	Food	*pH Range*
Dairy Products	Butter	6.1 – 6.4
	Buttermilk	4.5
	Milk	6.3 – 6.5
	Cream	6.5
	Cheese (American mild and cheddar)	4.9; 5.9
	Yogurt	3.8 – 4.2
Meat and Poultry	Beef (ground)	5.1 – 6.2
	Ham	5.9 – 6.1
	Veal	6.0
	Chicken	6.2 – 6.4
Fish and Shellfish	Fish (most species)	6.6 – 6.8
	Clams	6.5
	Crabs	7.0
	Oysters	4.8 – 6.3
	Tuna Fish	5.2 – 6.1
	Shrimp	6.8 – 7.0
	Salmon	6.1 – 6.3
	White Fish	5.5

Contd...

	Food	*pH Range*
Fruits and Vegetables	Apples	2.9 – 3.3
	Apple Cider	3.6 – 3.8
	Bananas	4.5 – 4.7
	Figs	4.6
	Grapefruit (juice)	3.0
	Limes	1.8 – 2.0
	Honeydew melons	6.3 – 6.7
	Oranges (juice)	3.6 – 4.3
	Plums	2.8 – 4.6
	Watermelons	5.2 – 5.6
	Grapes	3.4 – 4.5
	Asparagus (buds and stalks)	5.7 – 6.1
	Beans (string and lima)	4.6 – 6.5
	Beets (sugar)	4.2 - 4.4
	Broccoli	6.5
	Brussels Sprouts	6.3
	Cabbage (green)	5.4 – 6.0
	Carrots	4.9 – 5.2; 6.0
	Cauliflower	5.6
	Celery	5.7 – 6.0
	Corn (sweet)	7.3
	Cucumbers	3.8
	Eggplant	4.5
	Eggs yolks (white)	6.0 – 6.3 (7.6– 9.5)
	Lettuce	6.0
	Olives (green)	3.6 – 3.8
	Onions (red)	5.3 – 5.8
	Parsley	5.7 – 6.0
	Parsnip	5.3
	Potatoes (tubers and sweet)	5.3 – 5.6
	Pumpkin	4.8 – 5.2
	Rhubarb	3.1 – 3.4
	Spinach	5.5 – 6.0
	Squash	5.0 – 5.4
	Tomatoes (whole)	4.2 – 4.3
	Turnips	5.2 – 5.5

Table: Proportion of total acid undissociated at different pH values (expressed as percentages)

	pH Values				
Organic Acids	3	4	5	6	7
Acetic acid	98.5	84.5	34.9	5.1	0.54
Benzoic acid	93.5	59.3	12.8	1.44	0.144
Citric acid	53.0	18.9	0.41	0.006	<0.001
Lactic acid	86.6	39.2	6.05	0.64	0.064
Methyl, ethyl, propyl parabens	>99.99	99.99	99.96	99.66	96.72
Propionic acid	98.5	87.6	41.7	6.67	0.71
Sorbic acid	97.4	82.0	30.0	4.1	0.48

Nutrient Content

Microorganisms require certain basic nutrients for growth and maintenance of metabolic functions. The amount and type of nutrients required range widely depending on the microorganism. These nutrients include water, a source of energy, nitrogen, vitamins, and minerals.

Varying amounts of these nutrients are present in foods. Meats have abundant protein, lipids, minerals, and vitamins. Most muscle foods have low levels of carbohydrates. Plant foods have high concentrationsof different types of carbohydrates and varying levels of proteins, minerals, and vitamins. Foods such as milk and milk products and eggs are rich in nutrients. Foodborne microorganisms can derive energy from carbohydrates, alcohols, and amino acids. Most microorganisms will metabolise simple sugars such as glucose. Others can metabolise more complex carbohydrates, such as starch or cellulose found in plant foods, or glycogen found in muscle foods. Some microorganisms can use fats as an energy source.

Amino acids serve as a source of nitrogen and energy and are utilised by most microorganisms. Some microorganisms are able to metabolise peptides and more complex proteins.

Other sources of nitrogen include, for example, urea, ammonia, creatinine, and methylamines.

Examples of minerals required for microbial growth include phosphorus, iron, magnesium, sulphur, manganese, calcium, and potassium. In general, small amounts of these minerals are required; thus a wide range of foods can serve as good sources of minerals.

In general, the Gram (+) bacteria are more fastidious in their nutritional requirements and thus are not able to synthesise certain nutrients required for growth. For example, the Gram (+) foodborne pathogen *S. aureus* requires amino acids, thiamine, and nicotinic acid for growth. Fruits and vegetables that are deficient in B vitamins do not effectively support the growth of these microorganisms. The Gram (-) bacteria are generally able to derive their basic nutritional requirements from the existing carbohydrates, proteins, lipids, minerals, and vitamins that are found in a wide range of food.

An example of a pathogen with specific nutrient requirements is *Salmonella* Enteritidis. Growth of *Salmonella* Enteritidis may be limited by the availability of iron. For example, the albumen portion of the egg, as opposed to the yolk, includes anti-microbial agents and limited free iron that prevent the growth of *Salmonella* Enteritidis to high levels. Clay and Board (1991) demonstrated that the addition of iron to an inoculum of *Salmonella* Enteritidis in egg albumen resulted in growth of the pathogen to higher levels compared to levels reached when a control inoculum (without iron) was used.

The microorganisms that usually predominate in foods are those that can most easily utilise the nutrients present. Generally, the simple carbohydrates and amino acids are utilised first, followed by the more complex forms of these nutrients. The complexity of foods in general is such that several microorganisms can be growing in a food at the same time. The rate of growth is limited by the availability of essential nutrients.

Biological Structure

Plant and animal derived foods, especially in the raw state, have biological structures that may prevent the entry and growth of pathogenic microorganisms. Examples of such physical barriers include testa of seeds, skin of fruits and vegetables, shell of nuts, animal hide, egg cuticle, shell, and membranes.

Plant and animal foods may have pathogenic microorganisms attached to the surface or trapped within surface folds or crevices. Intact biological structures thus can be important in preventing entry and subsequent growth of microorganisms. Several factors may influence penetration of these barriers. The maturity of plant foods will influence the effectiveness of the protective barriers. Physical damage due to handling during harvest, transport, or storage, as well as invasion of insects can allow the penetration of microorganisms. During the preparation of foods, processes such as slicing, chopping, grinding, and shucking will destroy the physical barriers. Thus, the interior of the food can become contaminated and growth can occur depending on the intrinsic properties of the food. For example, *Salmonella* spp. have been shown to grow on the interior of portions of cut cantaloupe, watermelon, honeydew melons (Golden and others 1993), and tomatoes (Lin & Wei 1997), given sufficient time and temperature.

Fruits are an example of the potential of pathogenic microorganisms to penetrate intact barriers. After harvest, pathogens will survive but usually not grow on the outer surface of fresh fruits and vegetables.

Growth on intact surfaces is not common because foodborne pathogens do not produce the enzymes necessary to break down the protective outer barriers on most produce. This outer barrier restricts the availability of nutrients and moisture. One exception is the reported growth of *E. coli* O157:H7 on the surface of watermelon and cantaloupe rinds (del Rosario and Beuchat 1995). Survival of foodborne pathogens on produce is significantly enhanced once the protective epidermal barrier has been broken either by physical damage, such as punctures or bruising, or by degradation by plant pathogens (bacteria or fungi). These conditions can also promote the multiplication of pathogens, especially at higher temperatures. Infiltration of fruit was predicted and described by Bartz and Showalter (1981) based on the general gas law, which states that any change in pressure of an ideal gas in a closed container of constant volume is directly proportional to a change in temperature of the gas. In their work, Bartz and Showalter describe a tomato; however, any fruit, such as an apple, can be considered a container that is not completely closed.

As the container or fruit cools, the decrease in internal gas pressure results in a partial vacuum inside the fruit, which then results in an influx from the external environment. For example, an influx of pathogens from the fruit surface or cooling water could occur as a result of an increase in external pressure due to immersing warm fruit in cool water. Internalisation of bacteria into fruits and vegetables could also occur due to breaks in the tissues or through morphological structures in the fruit itself, such as the calyx or stem scar. Although infiltration was considered a possible scenario, the panel concluded that there is insufficient epidemiological evidence to require refrigeration of intact fruit.

The egg is another good example of an effective biological structure that, when intact, will prevent external microbial contamination of the perishable yolk; contamination is possible, however, through transovarian infection. For the interior of an egg to become contaminated by microorganisms on the surface, there must be penetration of the shell and its membranes. In addition, the egg white contains anti-microbial factors. When there are cracks through the inner membrane of the egg, microorganisms penetrate into the egg. Factors such as temperature of storage, relative humidity, age of eggs, and level of surface contamination will influence internalisation. For example, conditions such as high humidity and wet and dirty shells, along with a drop in the storage temperature will increase the likelihood for entry of bacteria. If eggs are washed, the wash water should be 12°C (22°F) higher than the temperature of the eggs to prevent microbial penetration. After washing, the eggs should be dried and then cooled. The Food and Drug Administration (FDA) published a final rule that applies to shell eggs that have not be processed to destroy all live Salmonella before distribution to the consumer. The rule mandates that eggs should be kept dry and chilled below 7.2°C (45°F) to prevent growth of *Salmonella* Enteritidis. Heating of food as well as other types of processing will break down protective biological structures and alter such factors as pH and a_w. These changes could potentially allow the growth of microbial pathogens.

Redox Potential

The oxidation-reduction or redox potential of a substance is defined in terms of the ratio of the total oxidising (electron accepting) power to the total reducing (electron donating) power of the substance. In effect, redox potential is a measurement of the ease by which a substance gains or loses electrons. The redox potential (Eh) is measured in terms of millivolts. A fully oxidised standard oxygen electrode will have an Eh of +810 mV at pH 7.0, 30°C (86°F), and under the same conditions, a completely reduced standard hydrogen electrode will have an Eh of –420 mV. The Eh is dependent on the pH of the substrate; normally the Eh is taken at pH 7.0.

The major groups of microorganisms based on their relationship to Eh for growth are aerobes, anaerobes, facultative aerobes, and microaerophiles. Examples of foodborne pathogens for each of these classifications include *Aeromonas hydrophila, Clostridium botulinum, Escherichia coli* O157: H7, and *Campylobacter jejuni*, respectively. Generally, the range at which different microorganisms can grow are as follows: aerobes +500 to +300 mV; facultative anaerobes +300 to –100 mV; and anaerobes +100 to less than –250 mV. For example, *C. botulinum* is a strict anaerobe that requires an Eh of less than +60 mV for growth; however, slower growth can occur at higher Eh values. The relationship of Eh to growth can be significantly affected by the presence of salt and other food constituents. For example, in one study with smoked herring, toxin was produced in inoculated product stored at 15°C (59°F) within three days at an Eh of +200 to +250 mV (Huss and others 1979). In this case, the major oxidant would be trimethylamine oxide, which becomes the electron acceptor for *C. botulinum*. The anaerobe *Clostridium perfringens* can initiate growth at an Eh close to +200 mV; however, in the presence of increasing concentrations of certain substances, such as salt, the limiting Eh increases (Morris 2000).

The measured Eh values of various foods depend on changes in the pH of the food, microbial growth, packaging, the partial pressure of oxygen in the storage environment, and ingredients and composition (protein, ascorbic acid, reducing sugars, oxidation level of cations, and so on). Another important factor is the poising capacity of the food. Poising capacity, which is analogous to buffering capacity, relates to the extent to which a food resists external affected changes in Eh. The poising capacity of the food will be affected by oxidising and reducing constituents in the food as well as by the presence of active respiratory enzyme systems. Fresh fruits and vegetables and muscle foods will continue to respire; thus low Eh values can result (Morris 2000).

The measurement of redox potential of food is done rather easily, either for single or multicomponent foods. For multicomponent foods, in addition to measurement of each component, the redox potential of the interface areas and microenvironments should be considered. However, difficulties arise in taking accurate measurements and in accounting for the differences throughout the food and the equilibrium at the point of measurement. According to Morris (2000): "This imposes the further requirements: (1) that the measuring electrode be so prepared and calibrated that it gives stable and reproducible readings, and (2) that a foodstuff is tested in a manner that does not cause any change in the potential that is to be measured. It would be unwise to use redox potential information in isolation to predict food safety, or to rely exclusively on control of redox potential as the means of preventing growth of specific microorganisms." Redox measurements could possibly be used in combination with other factors to evaluate the potential for pathogen growth. However, the limitations make it a rather difficult and variable factor that could result in erroneous conclusions in the absence of other comprehensive information.

Naturally Occurring and Added Anti-microbials

Some foods intrinsically contain naturally-occurring anti-microbial compounds that convey some level of microbiological stability to them. There are a number of plant-based anti-microbial constituents, including many essential oils, tannins, glycosides, and resins, that can be found in certain foods. Specific examples include eugenol in cloves, allicin in garlic, cinnamic aldehyde and eugenol in cinnamon, allyl isothiocyanate in mustard, eugenol and thymol in sage, and carvacrol (isothymol) and thymol in oregano. Other plant-derived anti-microbial constituents include the phytoalexins and the lectins. Lectins are proteins that can specifically bind to a variety of polysaccharides, including the glycoproteins of cell surfaces. Through this binding, lectins can exert a slight anti-microbial effect. The usual concentration of these compounds in formulated foods is relatively low, so that the anti-microbial effect alone is slight. However, these compounds may produce greater stability in combination with other factors in the formulation.

Some animal-based foods also contain anti-microbial constituents. Examples include lactoferrin, conglutinin and the lactoperoxidase system in cow's milk, lysozyme in eggs and milk, and other factors in fresh meat, poultry and seafood. Lysozyme is a small protein that can hydrolyse the cell wall of bacteria. The lactoperoxidase system in bovine milk consists of three distinct components that are required for its anti-microbial action: lactoperoxidase, thiocyanate, and hydrogen peroxide. Gram (-) psychotrophs such as the pseudomonads have been shown to be very sensitive to the lactoperoxidase system. Consequently, this system, in an enhanced form, has been suggested to improve the keeping quality of raw milk in developing countries where adequate refrigeration is scarce. Similar to the plant-derived anti-microbial compounds, the animal-derived compounds have a limited effect on ambient shelf life of foods.

It is also known that some types of food processing result in the formation of anti-microbial compounds in the food. The smoking of fish and meat can result in the deposition of anti-microbial substances onto the product surface. Maillard compounds resulting from condensation reactions between sugars and aminoacids or peptides upon heating of certain foods can impart some anti-microbial activity. Smoke condensate includes phenol, which is not only an anti-microbial, but also lowers the surface pH. Some processors also lower the surface pH with liquid smoke to achieve an unsliced shelf-stable product.

Some types of fermentations can result in the natural production of anti-microbial substances, including bacteriocins, antibiotics, and other related inhibitors. Bacteriocins are proteins or peptides that are produced by certain strains of bacteria that inactivate other, usually closely-related, bacteria. The most commonly characterised bacteriocins are those produced by the lactic acid bacteria.

The lantibiotic nisin produced by certain strains of *Lactococcus lactis* is one of the best characterised of the bacteriocins. Nisin is approved for food applications in over 50 countries around the world. Nisin's first food application was to prevent late-blowing in Swiss cheese by *Clostridium butyricum*. Nisin is a polypeptide that is effective against most Gram (+) bacteria but is ineffective against Gram (-) organisms and fungi. Nisin can be produced in the food by starter cultures or, more commonly, it can be used as an additive in the form of a standardised preparation. Nisin has been used to effectively control spore-forming organisms in processed cheese formulations, and has been shown to have an interactive effect with heat.

There are a number of other bacteriocins and natural anti-microbials that have been described, however, these have found very limited application in commercial use as food preservatives because of their restricted range of activity, limited compatibility with the food formulation or their regulatory status.

In addition to naturally-occurring anti-microbial compounds in foods, a variety of chemical preservatives and additives can extend the shelf life of food and/or inhibit pathogens, either singly or in combination.

The selection and use of these preservatives is typically governed by food law regulation of a country or region of the world. A number of criteria should be followed when selecting a preservative for a specific food application. Ideally, the preservative should have a wide spectrum of activity against the target spoilage organisms and pathogens expected to be encountered in the food. The preservative must be active for the desired shelf life of the food and under the expected formulation conditions in the food. It should cause minimal organoleptic impact on the food and should not interfere with desirable microbiological processes expected to occur in the food, such as the ripening of cheese or leavening of baked goods.

Added anti-microbial compounds can have an interactive or synergistic effect with other parameters of the formulation. One example is the interaction with pH. Many preservatives have an optimum pH range for effectiveness. Other factors include a_w, presence of other preservatives, types of food constituents, presence of certain enzymes, processing temperature, storage atmosphere, and partition coefficients. The effective use of combinations of preservatives with other physicochemical parameters of a food formulation can stabilise that food against spoilage organisms or pathogens. Leistner systematically developed the "hurdle concept" to describe these effects (Leistner 1995). The hurdle concept states that several inhibitory factors (hurdles), while individually unable to inhibit microorganisms, will, nevertheless, be effective in combination. A classic example of applying the hurdle concept is the anti-botulinal stability of certain shelf-stable processed cheese formulations.

Combinations of moisture, total salt, and pH have been shown to allow for the safe storage of these products at room temperature for extended time even though the individual factors, taken singly, would not support that practice. In combination products, the effectiveness of an anti-microbial may be altered by other factors including the potential for migration of the anti-microbial to other components of the food and the different food parameters at the interface areas. There are a number of food formulations that, either by addition of preservatives or through the application of the hurdle concept do not require refrigeration for microbiological stability or safety.

However, in the absence of a well-defined and validated microbiological model, it is usually difficult to evaluate the microbiological safety of these products. In the majority of these cases, the application of appropriate microbiological challenge testing is the most effective tool for judging the suitability of these formulations for non-refrigerated storage.

Competitive Microflora

The potential for microbial growth of pathogens in temperature-sensitive foods depends on the combination of the intrinsic and extrinsic factors, and the processing technologies that have been applied. Within the microbial flora in a food, there are many important biological attributes of individual organisms that influence the species that predominates. These include the individual growth rates of the microbial strains and the mutual interactions or influences among species in mixed populations.

Growth

In a food environment, an organism grows in a characteristic manner and at a characteristic rate. The length of the lag phase, generation time, and total cell yield are determined by genetic factors. Accumulation of metabolic products may limit the growth of particular species. If the limiting metabolic product can be used as a substrate by other species, these may take over (partly or wholly), creating an association or succession. Due to the complex of continuing interactions between environmental factors and microorganisms, a food at any one point in time has a characteristic flora, known as its association. The microbial profile changes continuously and one association succeeds another in what is called succession. Many examples of this phenomenon have been observed in the microbial deterioration and spoilage of foods.

As long as metabolically active organisms remain, they continue to interact, so that dominance in the flora occurs as a dynamic process. Based on their growth-enhancing or inhibiting nature, these interactions are either antagonistic or synergistic.

Competition

In food systems, antagonistic processes usually include competition for nutrients, competition for attachment/adhesion sites (space), unfavourable alterations of the environment, and a combination of these factors. Early studies demonstrated that the natural biota of frozen pot pies inhibited inoculated cells of *S. aureus, E. coli* and *Salmonella* Typhimurium. Another example of this phenomenon is raw ground beef. Even though *S. aureus* is often found in low numbers in this product, staphylococcal enterotoxin is not produced. The reason is that the *Pseudomonas-Acinetobacter-Moraxella* association that is always present in this food grows at a higher rate, outgrowing the staphylococci.

Organisms of high metabolic activity may consume

required nutrients, selectively reducing these substances, and inhibiting the growth of other organisms. Depletion of oxygen or accumulation of carbon dioxide favours facultative obligate anaerobes which occur in vacuum-packaged fresh meats held under refrigeration.

Staphylococci are particularly sensitive to nutrient depletion. Coliforms and *Pseudomonas* spp. may utilise amino acids necessary for staphylococcal growth and make them unavailable. Other genera of Micrococcaceae can utilise nutrients more rapidly than staphylococci. Streptococci inhibit staphylococci by exhausting the supply of nicotinamide or niacin and biotin. *Staphylococcus aureus* is a poor competitor in both fresh and frozen foods. At temperatures that favour staphylococcal growth, the normal food saprophytic biota offers protection against staphylococcal growth through antagonism, competition for nutrients, and modification of the environment to conditions less favourable to *S. aureus*. Changes in the composition of the food, as well as changes in intrinsic or extrinsic factors may either stimulate or decrease competitive effects.

Effects on Growth Inhibition

Changes in growth stimulation have been reported among several foodborne organisms, including yeasts, micrococci, streptococci, lactobacilli and Enterobacteriaceae. Growth stimulating mechanisms can have a significant influence on the buildup of a typical flora. There are several of these mechanisms, a few of which are listed below:

- Metabolic products from one organism can be absorbed and utilised by other organisms.
- Changes in pH may promote the growth of certain microorganisms. An example is natural fermentations, in which acid production establishes the dominance of acid tolerant organisms such as the lactic acid bacteria. Growth of moulds on high acid foods has been found to raise the pH, thus stimulating the growth of *C. botulinum*.

- Changes in Eh or a_w in the food can influence symbiosis. At warm temperatures, *C. perfringens* can lower the redox potential in the tissues of freshly slaughtered animals so that even more obligately anaerobic organisms can grow.
- There are some associations where maximum growth and normal metabolic activity are not developed unless both organisms are present.

This information can be used in the hurdle concept to control microorganisms in temperature-sensitive foods.

Extrinsic Factors

Types of Packaging/Atmospheres

Many scientific studies have demonstrated the anti-microbial activity of gases at ambient and sub-ambient pressures on microorganisms important in foods.

Gases inhibit microorganisms by two mechanisms. First, they can have a direct toxic effect that can inhibit growth and proliferation. Carbon dioxide (CO_2), ozone (O_3), and oxygen (O_2) are gases that are directly toxic to certain microorganisms. This inhibitory mechanism is dependent upon the chemical and physical properties of the gas and its interaction with the aqueous and lipid phases of the food. Oxidising radicals generated by O_3 and O_2 are highly toxic to anaerobic bacteria and can have an inhibitory effect on aerobes depending on their concentration. Carbon dioxide is effective against obligate aerobes and at high levels can deter other microorganisms.

A second inhibitory mechanism is achieved by modifying the gas composition, which has indirect inhibitory effects by altering the ecology of the microbial environment. When the atmosphere is altered, the competitive environment is also altered. Atmospheres that have a negative effect on the growth of one particular microorganism may promote the growth of another. This effect may have positive or negative consequences depending upon the native pathogenic microflora and their substrate. Nitrogen replacement of oxygen is an example of this indirect anti-microbial activity.

A variety of common technologies are used to inhibit the growth of microorganisms, and a majority of these methods rely upon temperature to augment the inhibitory effects. Technologies include modified atmosphere packing (MAP), controlled atmosphere packaging (CAP), controlled atmosphere storage (CAS), direct addition of carbon dioxide (DAC), and hypobaric storage.

Controlled atmosphere and modified atmosphere packaging of certain foods can dramatically extend their shelf life. The use of CO_2, N_2, and ethanol are examples of MAP applications. In general, the inhibitory effects of CO_2 increase with decreasing temperature due to the increased solubility of CO_2 at lower temperatures. Carbon dioxide dissolves in the food and lowers the pH of the food.

Nitrogen, being an inert gas, has no direct anti-microbial properties. It is typically used to displace oxygen in the food package either alone or in combination with CO_2, thus having an indirect inhibitory effect on aerobic microorganisms. The preservation principle of anti-microbial atmospheres has been applied to fruits and vegetables, raw beef, chicken and fish, dairy foods including milk and cottage cheese, eggs, and a variety of prepared, ready-to-eat foods. There are several intrinsic and extrinsic factors that influence the efficacy of anti-microbial atmospheres.

These factors-including product temperature, product-to-headspace gas volume ratio, initial microbial loads and type of flora, package barrier properties, and biochemical composition of the food-all interact to determine the degree to which the microbial quality and safety are enhanced. Temperature, the most important factor affecting the efficacy of anti-microbial atmospheres, directly affects growth rate, but also indirectly affects growth by affecting gas solubility. At practical food storage temperatures, packaging configurations, especially the product-to-headspace volume ratio, play a major role in determining the magnitude of microbial inhibition.

In MAP, package barrier properties have a major effect on the microbial growth by influencing the time in which the selected modified atmosphere gases remain in contact with the product and the rate at which oxygen enters the package.

Water activity, salt content of the aqueous phase, pH, and fat content of foods also play a role in overall inhibitory effects of anti-microbial gases. As with temperature, the physical and chemical characteristics of the food have an effect on the solubility of the inhibitory gas. For example, increasing salt concentrations decreases CO_2 solubility.

The major safety consideration in extending shelf life of foods by MAP or related technologies is the loss of sensory cues to spoilage provided by bacterial growth. Without spoilage bacteria indicators, it is conceivable that a food could have acceptable organoleptic quality, but be unsafe. The effect of loss of competitive inhibition by spoilage bacteria is most pronounced on the facultative anaerobic pathogenic bacterial populations in foods under altered atmospheres. By combining anti-microbial atmospheres with other techniques, hurdle technology strategies may be generated that can further enhance food quality and safety.

Effect of Time/Temperature Conditions on Microbial Growth

Impact of Time: When considering growth rates of microbial pathogens, in addition to temperature, time is a critical consideration. Food producers or manufacturers address the concept of time as it relates to microbial growth when a product's shelf life is determined. Shelf life is the time period from when the product is produced until the time it is intended to be consumed or used. Several factors are used to determine a product's shelf life, ranging from organoleptic qualities to microbiological safety. For the purpose of this report, the key consideration is the microbiological safety of the product. The Uniform Open Dating Regulation requires the shelf life of a perishable food product to be expressed in terms of a "sell by" date. The "sell by" date must incorporate the shelf life of the product plus a reasonable period for consumption that consists of at least one-third of the approximate total shelf life of the perishable food product.

At retail or foodservice, an additional period of time referred to herein as "use-period" should also be considered. As an example, fast food locations may find it operationally desirable to hold processed cheese slices at ambient temperatures for a complete shift or meal period, which may be in excess of 4 h.

This practice provides operational efficiency by allowing the cheese to melt faster on a hot sandwich as well as providing a better quality sandwich. Although refrigeration may be required for safety under long-term storage conditions, for use-periods measured in hours, storage at ambient temperatures may be acceptable. Under certain circumstances, time alone at ambient temperatures can be used to control product safety.

When time alone is used as a control, the duration should be equal to or less than the lag phase of the pathogen(s) of concern in the product in question. For refrigerated food products, the shelf life or use period required for safety may vary depending on the temperature at which the product is stored. For example, Mossel and Thomas (1988) report that the lag time for growth of *L. monocytogenes* at 10°C (50°F) is 1.5 d, while at 1°C (34°F) lag time is ~3.3 d. Likewise, they report that at 10°C (50°F) the generation time for the same organism is 5-8 h, while at 1°C (34°F), the generation time is between 62 and 131 h.

The data were obtained by using the USDA Pathogen Micromodel Programme at a NaCl concentration of 2 per cent and a_w of 0.989. It should be noted that this model was developed in broth under various salt and pH combinations, and that growth of bacteria in food systems will likely differ. According to the model results, a temperature shift from 10°C (50°F) to 25°C (77°F) decreases the lag time of *L. monocytogenes* from 60 to 10 h. In a similar manner, a pH increase from 4.5 to 6.5 decreases the lag time from 60 to 5 h. In conclusion, the safety of a product during its shelf life may differ, depending upon other conditions such as temperature of storage, pH of the product, and so on. This study by Mossel and Thomas (1988), along with numerous others, illustrates that various time/temperature combinations can be used to control product safety depending on the product's intended use.

Impact of Temperature

All microorganisms have a defined temperature range in which they grow, with a minimum, maximum, and optimum. An understanding of the interplay between time, temperature, and other intrinsic and extrinsic factors is crucial to selecting the proper storage conditions for a food product. Temperature has dramatic impact on both the generation time of an organism and its lag period. Over a defined temperature range, the growth rate of an organism is classically defined as an Arrhenius relationship. The log growth rate constant is found to be proportional to the reciprocal of the absolute temperature.

When temperatures approach the maxima for a specific microorganism, the growth rate declines more rapidly than when temperatures approach the minima for that same microorganism.

At low temperatures, two factors govern the point at which growth stops: (1) reaction rates for the individual enzymes in the organism become much slower, and (2) low temperatures reduce the fluidity of the cytoplasmic membrane, thus interfering with transport mechanisms. At high temperatures, structural cell components become denatured and inactivation of heat-sensitive enzymes occurs. While the growth rate increases with increasing temperature, the rate tends to decline rapidly thereafter, until the temperature maximum is reached.

The relationship between temperature and growth rate constant varies significantly across groups of microorganisms. Four major groups of microorganisms have been described based on their temperature ranges for growth: thermophiles, mesophiles, psychrophiles, and psychrotrophs. The optimum temperature for growth of thermophiles is between 55 to 65°C (131 to 149°F) with the maximum as high as 90°C (194°F) and a minimum of around 40°C (104°F). Mesophiles, which include virtually all human pathogens, have an optimum growth range of between 30°C (86°F) and 45°C (113°F), and a minimum growth temperature ranging from 5 to 10°C (41 to 50°F). Psychrophilic organisms have an optimum growth range of 12°C (54°F) to 15°C (59°F) with a maximum range of 15°C (59°F) to 20°C (68°F). There are very few true psychrophilic organisms of consequence to foods. Psychrotrophs such as *L. monocytogenes* and *C. botulinum* type E are capable of growing at low temperatures (minimum of - 0.4°C [31°F] and 3.3°C [38°F], respectively, to 5°C [41°F]), but have a higher growth optimum range (37°C [99°F] and 30°C [86°F], respectively) than true psychrophiles. Psychrotrophic organisms are much more relevant to food and include spoilage bacteria, spoilage yeast and moulds, as well as certain foodborne pathogens.

Growth temperature is known to regulate the expression of virulence genes in certain foodborne pathogens. For example, the expression of proteins governed by the *Yersinia enterocolitica* virulence plasmid is high at 37°C (99°F), low at 22°C (72°F), and not detectable at 4°C (39°F). Growth temperature also impacts an organism's thermal sensitivity. *Listeria monocytogenes*, when held at 48°C (118°F) in inoculated sausages, has an increase of 2.4-fold in its D.

It must be emphasised that the lag period and growth rate of a microorganism are influenced not only by temperature but by other intrinsic and extrinsic factors as well. For example, the growth rate of *Clostridium perfringens* is significantly lower at pH 5.8 versus pH 7.2 across a wide range of temperatures. Salmonellae do not grow at temperatures below 5.2°C (41°F). The intrinsic factors of the food product, however, have been shown to impact the ability of salmonellae to grow at low temperatures. *Salmonella* Senftenberg, *S.* Enteritidis, and

S. Manhattan were not able to grow in ham salad or custard held at 10°C (50°F), but were able to grow in chicken a la king held at 7°C (45°F).

Staphylococcus aureus has been shown to grow at temperatures as low as 7°C (45°F), but the lower limit for enterotoxin production has been shown to be 10°C (50°F). In general, toxin production below about 20°C (68°F) is slow. For example, in laboratory media at pH 7, the time to produce detectable levels of enterotoxin ranged from 78 – 98 h at 19°C (66°F) to 14 – 16 h at 26°C (79°F). Less favourable conditions, such as reduced pH, slowed enterotoxin production even further.

When measuring the suitability of holding a refrigerated food at room temperature for a period of time, consideration may be given to each factor independently. Doing so, however, ignores the potential to safely hold products for a period of time out of refrigeration based on interaction effects. Consideration of each relevant factor independently may lead to the conclusion that it is not a safe practice to do so, while, in reality, it is actually safe based on the interactive effects. The most appropriate method for evaluating such interactive effects is through a properly designed microbiological challenge study using relevant target microorganisms. Appropriate, validated predictive microbiological models may also be employed for this purpose. The use of challenge studies and/or predictive models can yield scientific data that supports holding a product with a certain formulation for a given time and temperature. It is incumbent upon the producer to have specific knowledge of the food formulation to generate valid scientific data.

Storage/Holding Conditions

This discussion of storage conditions will be limited to the storage/holding temperature, and the time/temperature involved in cooling of cooked items, and the relative humidity to which the food or packaging material may be exposed. Other factors that may be included as important considerations for storage, such as the effectiveness of the packaging material at conserving certain characteristics.

When considering growth rate of microbial pathogens, time and temperature are integral and must be considered together. Increases in storage and/or display temperature will decrease the shelf life of refrigerated foods since the higher the temperature, the more permissive conditions are for growth. At the same time, those foods that have been cooked or reheated and are served or held hot may require appropriate time/temperature control for safety. For example, the primary organism of concern for cooked meat and meat-containing products is *C. perfringens*. Illness symptoms are caused by ingestion of large numbers (greater than 108) of vegetative cells. The organism has an optimal growth range of 43-47°C (109-116°F) and a growth range of 12-50°C (54-122°F). Generation times as short as 8 min have been reported in certain foods under optimal conditions. Thus time/temperature management is essential for product safety.

The literature is replete with examples of outbreaks of foodborne illness that have resulted from cooling food too slowly, a practice that may permit growth of pathogenic bacteria. Of primary concern in this regard are the spore-forming pathogens that have relatively short lag times and the ability to grow rapidly and/or that may normally be present in large numbers. Organisms that possess such characteristics include *C. perfringens*, and *Bacillus cereus*. As with *C. perfringens*, foodborne illness caused by *B. cereus* is typically associated with consumption of food that has supported growth of the organism to relatively high numbers. The FDA "Bad Bug Book" notes that "The presence of large numbers of *B. cereus* (greater than 106 organisms/g) in a food is indicative of active growth and proliferation of the organism and is consistent with a potential hazard to health". In this case, the time and temperature (cooling rate) of certain foods must be addressed to assure rapid cooling for safety.

The effect of the relative humidity of the storage

environment on the safety of foods is somewhat more nebulous. The effect may or may not alter the a_w of the food. Such changes are product dependent. The earlier discussion on a_w and its effect on microorganisms in foods provides some background information. In addition, the possibility of surface evaporation or condensation of moisture on a surface should be considered.

Generally, foods that depend on a certain a_w for safety or shelf life considerations will need to be stored such that the environment does not markedly change this characteristic. Foods will eventually come to moisture equilibrium with their surroundings. Thus, processors and distributors need to provide for appropriate storage conditions to account for this fact.

Packaging will play a major role in the vulnerability of the food to the influence of relative humidity. But even within a sealed container, moisture migration and the phenomenon of environmental temperature fluctuation may play a role. It has been observed that certain foods with low a_w can be subject to moisture condensing on the surface due to wide environmental temperature shifts. This surface water will result in microenvironments favourable to growth of spoilage, and possibly pathogenic, microorganisms. As a general guideline, the product should be held such that environmental moisture, including that within the package, does not have an opportunity to alter the a_w of the product in an unfavourable way.

Processing Steps

The current definition of "potentially hazardous foods" considers the effect of processing in much the same way that it considers pH and a_w: it divides foods into two categories. Low-acid canned foods in a hermetically-sealed container do not require temperature control for safety. This rigid definition fails to address less processed foods, in less robust packaging, which still would not require temperature control for safety. Consider a baked product, such as a pie, with a pH of 5.5 and a_w of 0.96. Since this product is baked to an internal temperature

>180°F (82°C) to set the product structure of the pie, it will not contain any viable vegetative pathogens. Any pathogenic spores that survive the baking process will be inhibited by the pH and a_w values. If the product is cooled and packaged under conditions that do not allow recontamination with vegetative pathogens, the product is safe and stable at room temperature until consumed, or until quality considerations (that is, staling) make it unpalatable.

Scientifically sound criteria for determining whether foods require time/temperature control for safety should consider: 1) processes that destroy vegetative cells but not spores (when product formulation is capable of inhibiting spore germination); 2) post-process handling and packaging conditions that prevent reintroduction of vegetative pathogens onto or into the product before packaging; and 3) the use of packaging materials that while they do not provide a hermetic seal, do prevent reintroduction of vegetative pathogens into the product.

Other Factors

Intended End-use of Product: In addition to carefully assessing how the product is produced and distributed, it is important to consider how the food will ultimately be prepared, handled, and/or stored by the end user. A food product that does not require time/temperature control for safety at one point in the food production or distribution chain may require time/temperature control at another point, depending on its intended use. For example, a thermally processed food that is hot-filled into its final packaging may not require refrigeration if spore-forming pathogens are not capable of outgrowth. However, once the food item is taken out of its original packaging, it may require time/temperature control for safety if the product is likely to be recontaminated during its intended use.

Product History and Traditional Use

The panel struggled with the concept of product history and traditional use as a means to determine the need for time/

temperature control for safety. For example, there are foods which have a long history of safe storage use at ambient temperatures, yet have formulations, pH, and a_w that would designate them as "temperature controlled for safety" (TCS) foods. Paramount among them is white bread, but products such as intact fruits and vegetables, other breads, bottled waters, and some processed cheeses have a history of being stored and used at ambient temperatures with no public health impact. In addition, moisture protein ratios (MPR) for shelf-stable fermented sausages were developed to ensure process control values for these sausages that also have a traditional history of safety as a non-TCS food.

Moreover, an evaluation of the food characteristics provides a scientific explanation for the products to be safely stored at ambient temperatures. For example, baking of bread controls the growth of pathogens in the interior, and the low a_w precludes the growth of pathogens on the outer surface, so that it can be stored safely at ambient temperatures. Clearly these products' traditional uses and histories provide a valid justification for a decision to be made based on history. Care must be observed, however, as this traditional history can be influenced by the intrinsic and extrinsic factors and any changes in product end-use, processes, formulation, physical structure, processing, distribution, and/or storage. Changes in any of these parameters may invalidate the sole use of history as a basis for decisions on whether a food needs temperature control for safety.

The panel recognises that the use of history as a factor to decide whether a product needs time/temperature control for safety can be subjective. As a guidance, one should determine whether the food in question or any of its ingredients have been previously implicated as a common vehicle of foodborne disease as a result of abuse or storage at ambient temperature. Of particular importance are the microbiological agents that may be of concern based on food formulation, or that may be responsible for illnesses associated with the food and the reported contributing factors that have led to documented illnesses. Has adequate temperature control been clearly documented as a factor that can prevent or reduce the risk of illness associated with the food? As intrinsic or extrinsic factors change (for example, MAP or greatly extended shelf life), historical evidence alone may not be appropriate in determining potential risk. Therefore, for a product to be identified as non-TCS based on history and traditional use, the intrinsic and extrinsic factors affecting microbial growth need to have remained constant. Lastly, product history alone should not be used as the sole factor in determining whether a food needs time/temperature control for safety.

Interactions of Factors

Traditional food preservation techniques have used combinations of pH, a_w, atmosphere, numerous preservatives, and other inhibitory factors. Microbiologists have often referred to this phenomenon as the "hurdle effect". For example, certain processed meat products and pickles may use the salt-to-moisture ratio (brine ratio) to control pathogens. USDA recognises this strategy in designating as shelf-stable semi-dry sausages with a moisture-protein ratio of less than or equal to 3.1:1 and pH less than or equal to 5.0.

In salad dressings and mayonnaise-type products, the acid-to-moisture ratio along with pH is the governing factor for pathogen control. An acid:moisture ratio > 0.70 in combination with a pH < 4.1 is often used as the pathogen-control target level for these products. Usually, these ratios are combined with other factors such as pH or added anti-microbials to effect pathogen control. It is the interaction of these factors that controls the ability of pathogens to proliferate in foods.

Despite this long-standing recognition of the concept of hurdle technology (the possible synergistic effect of combining different inhibitory factors), the current definition of potentially hazardous foods only considers pH and a_w independently, and does not address their interaction. The panel believes that these interactions have to be taken into consideration.

Scientific advances in predictive food microbiology over the last two decades have repeatedly shown that different inhibitory factors that might not prevent pathogen growth when considered singly will prevent pathogen growth when used in concert. USDA Pathogen Modelling Programme model was developed in broth with salt and pH combinations. Also, the salt used to control the aw results in additional microbial inhibitory effects that may be lacking if other compounds are used. The values are the time in hours needed for a 3 log increase in *S. aureus* concentration as a function of the pH and a_w values.

It is clear from the numerical values that even though a food might have a pH of 5.0 and an aw of 0.92 (for example), after 72 h at room temperature, it may show a minimal increase in *S. aureus* concentration, and thus not constitute a significant risk to public health.

9

Metagenomics

Metagenomics (also Environmental Genomics, Ecogenomics or Community Genomics) is the study of genomes recovered from environmental samples as opposed to from clonal cultures.

This relatively new field of genetic research allows the genomic study of organisms that are not easily cultured in a laboratory. Early molecular work in the field was conducted by Norman R. Pace and colleagues, who used PCR to explore the diversity of ribosomal RNA sequences from organisms present in uncultured environmental samples. Considerable efforts ensured that these were not PCR false positives and supported the existence of a complex community of unexplored species. Although this methodology was limited to exploring highly conserved, non-protein coding genes, it did support early microbial morphology-based observations that diversity was far more complex than was known by culturing methods.

The insights gained from these breakthrough studies led Pace to propose the idea of cloning DNA directly from environmental samples as early as 1985. This led to the first report of isolating and cloning bulk DNA from an environmental sample, published by Pace and colleagues in

1991 (J. Bacteriol. 173: 4371) while Pace was in the Department of Biology at Indiana University. Soon after that, Healy reported the metagenomic isolation of functional genes from "zoolibraries" constructed from a complex culture of environmental organisms grown in the laboratory on dried grasses in 1995. After leaving the Pace laboratory, Ed DeLong continued in the field and has published work that has largely laid the groundwork for contemporary metagenomics based on signature 16S sequences, beginning with his group's construction of libraries from marine samples (1996 J. Bacteriol. 178:591) in collaboration with scientists from Diversa (formerly Recombinant BioCatalysis, Inc.). Construction of libraries from soil DNA faced many technical challenges before the first successes were reported from the laboratories of Radomski (USPN 5,849.491 filed in 1995), Handelsman, Pernodet, and Gottschalk.

Metagenomics for isolation of expressed gene and gene pathways products (functional screening of libraries derived directly from environmental samples, without culturing) was first conceived and practised by Jay M. Short, head of research and a founder of Diversa Corporation, at least as early as 1994. Short published articles, and filed and was issued many US patents on the discovery of genes, gene pathways and their products from gene libraries made from communities of uncultured organisms, using a variety of functional and sequence based methods for screening. Following earlier patent filings, one of the earlier publications related to Short's inventions appeared in SIM News (46:3-8) in 1996 and was titled "The Discovery of New Biocatalysts from Microbial Diversity". There are over a dozen publications and patents between 1996 and 2005. This work by Short and co-workers extended substantially the analysis beyond 16S sequences and was a precursor to functional genome-wide analysis such as that of Tringe *et al.* (2005).

The term "metagenomics" was first used by Jo Handelsman and others in the University of Wisconsin Department of Plant

Pathology, and appeared in publication in 1998. Kevin Chen and Lior Pachter (researchers at the University of California, Berkeley) defined metagenomics as "the application of modern genomics techniques to the study of communities of microbial organisms directly in their natural environments, bypassing the need for isolation and lab cultivation of individual species."

Environmental Samples

Conventional sequencing begins with a culture of identical cells as a source of DNA. However early metagenomic studies revealed that there are probably large groups of microorganisms in many environments that cannot be cultured and thus cannot be sequenced. These early studies focused on 16S ribosomal RNA sequences which are relatively short, often conserved within a species, and generally different between species. Many 16S rRNA sequences have been found which do not belong to any known cultured species, indicating that there are numerous unisolated organisms out there.

Recovery of DNA sequences longer than a few thousand base pairs from environmental samples was very difficult until recent advances in molecular biological techniques, particularly related to constructing libraries in bacterial artificial chromosomes (BACs), provided better vectors for molecular cloning. In addition, advances in bioinformatics, refinements of DNA amplification, and proliferation of computational power have greatly aided the analysis of DNA sequences recovered from environmental samples by metagenomics.

In 2002, Mya Breitbart, Forest Rohwer, and colleagues used metagenomic to show that 200 litres of seawater contains over 5000 different viruses. Subsequent studies showed that there are >1000 viral species in human stool and possibly a million different viruses per kilogram of marine sediment, including many bacteriophages. Essentially all of the viruses in these studies were new species. A 2004 metagenomic study of the Sargasso Sea found DNA from nearly 2000 different species including 148 types of bacteria never seen before.

Another study, also from 2004, revealed the genomes of bacteria and archaea from an acid mine drainage system that had resisted attempts to culture them.

Because the collection of DNA from an environment is largely uncontrolled, large samples, often sometimes prohibitively so, are needed to fully resolve the genomes of underrepresented members of a microbial community. On the other hand, many such underrepresented organisms might never be noticed without metagenomic analysis if they are difficult to isolate using traditional culturing techniques.

Community Metabolism

Many bacterial communities show significant division of labour in metabolism. Waste products of some organisms are metabolites for others. Working together they turn raw resources into fully metabolised waste. Using comparative gene studies and expression experiments with microarrays or proteomics researchers can piece together a metabolic network that goes beyond species boundaries. Such studies require detailed knowledge about which versions of which proteins are coded by which species and even by which strains of which species. So, community genomic information is fundamental to the study of how metabolites move through a community to be processed.

Environmental Shotgun Sequencing: Since their discovery in the 1670s by Anton van Leeuwenhoek, an incredible amount has been learned about microorganisms and their importance to human health, agriculture, industry, ecosystem functioning, global biogeochemical cycles, and the origin and evolution of life. Nevertheless, it is what is not known that is most astonishing. For example, though there are certainly at least 10 million species of bacteria, only a few thousand have been formally described. This contrasts with the more than 350,000 described species of beetles. This is one of many examples indicative of the general difficulties encountered in studying organisms that we cannot readily see or collect in large samples for future analyses. It is thus not surprising that most major advances in microbiology can be traced to methodological advances rather than scientific discoveries per se.

Examples of these key revolutionary methods include the use of microscopes to view microbial cells, the growth of single types of organisms in the lab in isolation from other types (culturing), the comparison of ribosomal RNA (rRNA) genes to construct the first tree of life that included microbes, the use of the polymerase chain reaction (PCR) to clone rRNA genes from organisms without culturing them, and the use of high-throughput "shotgun" methods to sequence the genomes of cultured species. We are now in the midst of another such revolution—this one driven by the use of genome sequencing methods to study microbes directly in their natural habitats, an approach known as metagenomics, environmental genomics, or community genomics.

Typing and Counting Microbes in the Environment

One of the most important and conceptually straightforward steps in studying any ecosystem involves cataloguing the types of organisms and the numbers of each type. For a long time, such typing and counting was an almost insurmountable problem in microbiology. This is largely because physical appearance does not provide a valid taxonomic picture in microbes. Appearance evolves so rapidly that two closely related taxa may look wildly different and two distantly related taxa may look the same. This vexing problem was partially overcome in the 1980s through the use of rRNA-PCR. This method allows microorganisms in a sample to be phylogenetically typed and counted based on the sequence of their rRNA genes, genes that are present in all cell-based organisms. In essence, a database of rRNA sequences from known organisms functions like a bird field guide, and finding a rRNA-PCR product is akin to seeing a bird through binoculars. Rather than counting species, this approach focuses on "phylotypes," which are defined as organisms whose rRNA sequences are very similar to each other (a cutoff of >97 per cent or >99 per cent identical is frequently used). The ability to use phylotyping to determine who was out there in any microbial sample has revolutionised environmental microbiology, led to many discoveries, and convinced many people (myself included) to become microbiologists.

The selective targeting of a single gene makes rRNA-PCR an efficient method for deep community sampling. However, this efficiency comes with limitations, most of which are complemented or circumvented by the randomness and breadth of ESS. For example, examination of the random samples of rRNA sequences obtained through ESS has already led to the discovery of new taxa—taxa that were completely missed by PCR because of its inability to sample all taxa equally well (e.g.,). In addition, ESS provides the first robust sampling of genes other than rRNA, and many of these genes can be more useful for some aspects of typing and counting.

Some universal protein coding genes are better than rRNA both for distinguishing closely related strains (because of third position variation in codons) and for estimating numbers of individuals (because they vary less in copy number between species than do rRNA genes). Perhaps most significantly, ESS is providing groundbreaking insights into the diversity of viruses, which lack rRNA genes and thus were left out of the previous revolution.

Certainly, many challenges remain before we can fully realise the potential of ESS for the typing and counting of species, including making automated yet accurate phylogenetic trees of every gene, determining which genes are most useful for which taxa, combining data from different genes even when we do not know if they come from the same organisms, building up databases of genes other than rRNA, and making up for the lack of depth of sampling. If these challenges are met, ESS has the potential to rewrite much of what we thought we knew about the phylogenetic diversity of microbial life.

Approaches to Understanding Functions in Communities

A community is, of course, more than a list of types of organisms. One approach to understanding the properties and functioning of a microbial community is to start with studies of the different types of organisms and build up from these individuals to the community. Ideally, to do this one would culture each of the phylotypes and study its properties in the lab. Unfortunately, many, if not most, key microbes have not yet been cultured. Thus, for many years, the only alternative was to make predictions about the biology of particular phylotypes based on what was known about related organisms. Unfortunately, this too does not work well for microbes since very closely related organisms frequently have major biological differences. For example, *Escherichia coli* K12 and *E. coli* O157:H7 are strains of the same species (and considered to be the same phylotype), with genomes containing only about 4,000 genes, yet each possesses hundreds of functionally important genes not seen in the other strain. Such differences are routine in microbes, and thus one cannot make any useful inferences about what particular phylotypes are doing (e.g., type of metabolism, growth properties, role in nutrient cycling, or pathogenicity) based on the activities of their relatives.

These difficulties—the inability to culture most microbes and the functional disparities between close relatives—led to one of the first kinds of metagenomic analyses, wherein predictions of function were made from analysis of the sequence of large DNA fragments from representatives of known phylotypes. This approach has provided some stunning insights, such as the discovery of a novel form of phototrophy in the oceans. However, this large insert approach has the same limitation as predicting properties from characterised relatives—a single cell cannot possibly represent the biological functions of all members of a phylotype.

ESS provides an alternative, more global way of assessing biological functions in microbial communities. As when using the large insert approach, functions can be predicted from sequences. However, in this case the predicted functions represent a random sampling of those encoded in the genomes of all the organisms present. This approach has unquestionably been wildly successful in terms of gene discovery. For example, analysis of ESS data has revealed novel forms of every type of gene family examined, as well as a great number of completely novel families. However, there is a major caveat when using ESS data to make community-level inferences. Ecosystems are more than just a bag of genes—they are made up of compartments (e.g., cells, chromosomes, and species), and these compartments matter. The key challenge in analysing ESS data is to sort the DNA fragments (which are usually less than 1,000 base pairs long relative to genome sizes of millions or billions of bases) into bins that correspond to compartments in the system being studied.

A recent study illustrates the importance of compartments when interpreting ESS data. When we analysed ESS data from symbionts living inside the gut of the glassy-winged sharpshooter (an insect that has a nutrient-limited diet), we were able to bin the data to two distinct symbionts. We then could infer from those data that one of the symbionts synthesises amino acids for the host while the other synthesises the needed vitamins and cofactors. Modelling and understanding of this ecosystem are greatly enhanced by the demonstration of this complementary division of labour, in comparison to simply knowing that amino acids, vitamins, and cofactors are made by "symbionts."

In considering the different binning methods and their limitations, the first question one needs to ask is, what are we trying to bin? Is it fragments from the same chromosome from a single cell, which would be useful for studying chromosome structure? If so, then perhaps genome assembly methods are the best. What if instead, as in the sharpshooter example, we are trying to have each bin include every fragment that came from a particular species, knowledge which may be useful for predicting community metabolic potential? If the level of genetic polymorphism among individual cells from the same species is high, then genome assembly methods may not work well (the polymorphisms will break up assemblies). A better approach might be to look for species-specific "word" frequencies in the DNA, such as ones created by patterns in codon usage. The challenge is, how do we tune the methods to find the right target level of resolution? If we are too stringent, most bins will include only a few fragments. But if we are too relaxed, we will create artificial constructs that may prove biologically misleading, such as grouping together sequences from different species. To make matters more complex, most likely the stringency needed will vary for different taxa present in the sample.

Methods of Binning

Another critical issue is the diversity of the system under study. Generally, binning works better when there are few different phylotypes present, all of which are distantly related and form discrete populations. This is why binning works well for the sharpshooter system and other relatively isolated, low diversity environments. Binning increases in difficulty exponentially as the number of species increases: the populations and species start to merge together, and the populations get more and more polymorphic and variable in relative abundance. Further complicating binning is the phenomenon of lateral gene transfer, where genes are exchanged between distantly related lineages at rates that are high enough that random sampling of a genome will frequently include genes with multiple histories.

Comparative Metagenomics

However, the area with perhaps the most promise involves the comparative analysis of different samples. This work parallels the comparative analysis of genomes of cultured species. Initial studies of that type compared distantly related taxa with enormous biological differences. What has been learned from these studies pertains mostly to core housekeeping functions, such as translation and DNA metabolism, and to other very ancient processes. It was not until comparisons were made between closely related organisms that we began to understand events that occurred on shorter time scales, such as selection, gene transfer, and mutation processes. Similarly, the initial comparisons of ESS data involved comparisons of wildly different environments, yielding insights into the general structure of communities. But as more comparisons are made between similar communities, such as those sampled during vertical and horizontal ocean transects, we will begin to learn about shorter time scale processes such as migration, speciation, extinction, responses to disturbance, and succession. It is from a combination of both approaches—comparing both similar and very divergent communities—that we will be able to understand the fundamental rules of microbial ecology and how they relate to ecological principles seen in macro-organisms.

It is helpful in this respect to compare ESS to the Internet. As with the Internet, ESS is a global portal for looking at what occurs in a previously hidden world. Making sense of it requires one to sort through massive, random, fragmented collections of bits of information. Such searches need to be done with caution because any time you analyse such a large amount of data patterns can be found. In addition, as with the Internet, there is certainly some hype associated with ESS that gives relatively trivial findings more attention than they deserve. As long as we treat ESS as a strong complement to existing methods, and we build the tools and databases necessary for people to use the information, it will live up to its revolutionary potential.

EBPR Metagenomics: The Solution to Pollution is Biotechnological Revolution

With the recent application of metagenomic techniques to wastewater treatment, the "black box" of enhanced biological phosphorus removal (EBPR) has been opened. Dirty limericks are included to illustrate some key points.

Wastewater is an age-old problem, as is crudely illustrated in the preceding limerick. Because of the adverse effects that wastewater is known to have on human health, the obvious solution is to get wastewater as far away from humans as possible. The ocean is a straightforward choice for wastewater disposal, and has been used for this purpose since antiquity. Modern wastewater engineering still rests on this paradigm, as illustrated by the engineer's tool of choice for measuring wastewater strength: the five-day BOD test. This test involves measuring how much oxygen is used up by a given sample of wastewater over a five day period. The rationale behind this test is that if wastewater is too strong (i.e. it contains a lot of biodegradable nutrients), and if it is released into too small a body of water, it will cause eutrophication. Eutrophication occurs when the algae in said body of water are overfed, in which case they'll not only consume all of the available nutrients, but in the process will consume all of the oxygen. The ensuing oxygen shortage is, of course, grievously problematic to the remainder of the aquatic ecosystem.

The five-day aspect of the five-day BOD test is hardly arbitrary. It was developed in Britain, where according to engineering lore any wastewater released into a stream or river will travel for a maximum of five days before reaching the ocean. Since the ocean represents such an enormous dilution factor, the wastewater is now effectively neutralised, particularly on paper. This "solution to pollution is dilution" method of wastewater disposal is still *de rigeur* in many parts of the world.

With an ever growing global population, and the realisation that water systems are not as limitless as was once believed, more stringent wastewater controls are coming into effect. The following limerick illustrates a more contemporary situation:

Certainly the protagonist of these preceding verses would appreciate an attempt to remedy eutrophication.

One of the main constituents of wastewater is phosphorus, which originates mostly (75-85 per cent) in detergents and excreta. Phosphorus is a very important nutrient in biological systems, but because it is a limiting nutrient in most freshwater systems, it is often a culprit in eutrophication. In order to prevent eutrophication, wastewater treatment plants employ processes to remove phosphorus before it is released into the environment. This can be done either through chemical precipitation with iron and aluminum salts, or through biological means termed enhanced biological phosphorus removal, or EBPR. The latter is more environmentally friendly, and has been implemented for over 30 years. One issue has provided a caveat to EBPR as a eutrophication panacea: the process occasionally and unpredictably fails.

Attempts to study the process of enhanced biological phosphorus removal from wastewater have been stymied by technological shortcomings. Although a model for the biochemical operation of EBPR is well characterised, the identity and metabolism of the participating organisms has remained obscure, rendering the EBPR sludge a "black box". A culture-based approach to studying the process produced a red herring that persisted for thirty years: the misidentification of the causative EBPR microbe as *Acinetobacter*. Recently, a new microbe has been identified as responsible for EBPR using molecular methods, and has been dubbed Candidatus *Accumulibacter phosphatis*. The microbe as yet defies attempts at isolation, and thus any culture-based investigations of its metabolism are impossible. October of 2006 saw the publication of a high-profile study that brought the tools of metagenomics to bear on this problem, paving the way for culture-independent study of *A. phosphatis'* metabolism, as well as the possibility of its cultivation. The function of *A. phosphatis* is illustrated in the following limerick:

Simply put, metagenomics is the study of entire genomes, isolated from environmental samples. The four-step process of constructing a metagenomic library is explained well at the SCQ, so suffice to say that this produces a lot of data.

Until recently metagenomic work was impossible, not

because metagenomics is an anagram for "commie agents", but rather due to technological limits. There simply wasn't enough raw computing power available to process the datasets that the metagenomic approach required, and it was very time consuming to sequence entire genomes. The advent of high-thorough put (lots of data produced quickly) techniques changed this, particularly the shotgun sequencing approach used for, among other things, Craig Venter's famous bid to sequence the human genome, and his application of the same technique to the microbial population of the Sargasso Sea. The shotgun sequencing approach is described at the SCQ: this same technique was used by Martin *et al.* to examine the populations of two EBPR systems, with the goal of sequencing the genome of *A. phosphates* so that its metabolism could be better understood.

Sequencing a genome is conceptually straightforward: extract the DNA from a sample, create a metagenomic library, sequence it with simple diagrams at the SCQ, and bingo! A dataset of code that you can now compare to databases from other sources, to see what genes you have. There are, however some intricacies to sequencing a genome from an environmental sample that bear exposition. First of all, if you have a bunch of environmental DNA, how can you tell which organism in the sample it came from?

In order to answer this question, it is first important to figure out what organisms are present in the sample. rRNA provides a handy tool for this step. rRNA is highly conserved across species, because the function of the ribosome is crucial to survival and highly specific. Thus the seqence of an organism's rRNA genes provide a unique "fingerprint" which permit its classification. This technique can be considered crude, but is effective for environmental samples given the magnitude of different species found in each sample. In the Martin *et al.* study, this technique was used to determine the community structure of the EBPR sludge, showing that *A. phosphates* was indeed present in the samples. These same genes served as targets for Fluorescent *In situ* Hybridisation (FISH) probing, wherein labelled probes specific for known *A. phosphates* rRNA genes, were used to locate the *A. phosphates* cells in the samples and show that they constituted the majority of the biomass of the EBPR sludge.

Shotgun sequencing does not produce a continuous genome sequence, even once all of the pieces have been laboriously put back together by computer. There remain gaps in the sequence. In the Martin *et al.* study, a fortuitous community structure aided in the reconstruction of the *A. phosphatis* genome: the only species present in both of the two different sludges sampled for the study was *A. phosphatis*. This allowed for a technique called "overlap binning" to be used, where any significant amount of overlap between sequenced DNA was assumed to be from *A. phosphatis*. This enabled the researchers to produce a more complete genome than would have been otherwise possible.

The resulting genomic data was compared with a huge resource for interpreting the genomic data: the existing genomes at the US Department of Energy Joint Genome Institute. The comparison allowed the researchers to see what genes *A. phosphatis* possessed, which allowed them to infer what enzymes the microbe was able to produce. This data in turn was used to reconstruct the metabolism of the microbe vis-à-vis EBPR, which the reader will recall was previously impossible.

The metagenomic work by Martin *et al.* has laid the groundwork for a new round of experimentation to optimise the EBPR process. In particular, the ability of *A. phosphatis* to use various carbon substrates is of interest. The majority of EBPR studies take place in small laboratory scale reactors, which are fed simple substrates such as acetate or propionate as they were in the Martin *et al.* study. This stands in contrast to full-scale EBPR plants, which receive a complex substrate: wastewater. The newly unravelled metabolic capabilities of *A. phosphatis* have also suggested a mechanism by which the microbe could be isolated and grown in pure culture. This work remains to be completed.

Armed with a better understanding of what makes the EBPR microbe tick, wastewater engineers can now begin to experiment with preventing EBPR failures, using metabolic data, rather than trial and error.

Humanity is moving from a paradigm which views the planet's capacity to absorb waste as limitless, exemplified by the five-day BOD test, to one where humans must create closed-loop systems to shield the environment from harm. Processes such as EBPR are crucial to this transition, and biotechnology in turn is crucial to the understanding of these biological processes.

10

Bioremediation

Bioremediation can be defined as any process that uses microorganisms, fungi, green plants or their enzymes to return the environment altered by contaminants to its original condition. Bioremediation may be employed to attack specific soil contaminants, such as degradation of chlorinated hydrocarbons by bacteria. An example of a more general approach is the cleanup of oil spills by the addition of nitrate and/or sulphate fertilizers to facilitate the decomposition of crude oil by indigenous or exogenous bacteria.

Naturally-occurring bioremediation and phytoremediation have been used for centuries. For example, desalination of agricultural land by phytoextraction has a long tradition. Bioremediation technology using microorganisms was reportedly invented by George M. Robinson. He was the assistant county petroleum engineer for Santa Maria, California. During the 1960s, he spent his spare time experimenting with dirty jars and various mixes of microbes.

Bioremediation technologies can be generally classified as *in situ* or *ex situ*. *In situ* bioremediation involves treating the contaminated material at the site while *ex situ* involves the removal of the contaminated material to be treated elsewhere. Some examples of bioremediation technologies are bioventing, landfarming, bioreactor, composting, bioaugmentation, rhizofiltration, and biostimulation.

Not all contaminants, however, are easily treated by bioremediation using microorganisms. For example, heavy metals such as cadmium and lead are not readily absorbed or captured by organisms. The assimilation of metals such as mercury into the food chain may worsen matters. Phytoremediation is useful in these circumstances, as many plants are able to bioaccumulate these toxins in their above-ground parts, which are then harvested for removal. The heavy metals in the harvested biomass may be further concentrated by incineration.

The quality of life on Earth is linked inextricably to the overall quality of the environment. In early times, we believed that we had an unlimited abundance of land and resources; today, however, the resources in the world show, in greater or lesser degree, our carelessness and negligence in using them.

The problems associated with contaminated sites now assume increasing prominence in many countries. Contaminated lands generally result from past industrial activities when awareness of the health and environmental effects connected with the production, use, and disposal of hazardous substances were less well recognised than today. The problem is worldwide, and the estimated number of contaminated sites is significant. It is now widely recognised that contaminated land is a potential threat to human health, and its continual discovery over recent years has led to international efforts to remedy many of these sites, either as a response to the risk of adverse health or environmental effects caused by contamination or to enable the site to be redeveloped for use.

The conventional techniques used for remediation have been to dig up contaminated soil and remove it to a landfill, or to cap and contain the contaminated areas of a site. The methods have some drawbacks. The first method simply moves the contamination elsewhere and may create significant risks in the excavation, handling, and transport of hazardous material. Additionally, it is very difficult and increasingly expensive to find new landfill sites for the final disposal of the material. The cap and contain method is only an interim solution since the contamination remains on site, requiring monitoring and maintenance of the isolation barriers long into the future, with all the associated costs and potential liability.

A better approach than these traditional methods is to completely destroy the pollutants if possible, or at least to transform them to innocuous substances. Some technologies that have been used are high-temperature incineration and various types of chemical decomposition (e.g., base-catalysed dechlorination, UV oxidation). They can be very effective at reducing levels of a range of contaminants, but have several drawbacks, principally their technological complexity, the cost for small-scale application, and the lack of public acceptance, especially for incineration that may increase the exposure to contaminants for both the workers at the site and nearby residents.

Bioremediation is an option that offers the possibility to destroy or render harmless various contaminants using natural biological activity. As such, it uses relatively low-cost, low-technology techniques, which generally have a high public acceptance and can often be carried out on site. It will not always be suitable, however, as the range of contaminants on which it is effective is limited, the time scales involved are relatively long, and the residual contaminant levels achievable may not always be appropriate. Although the methodologies employed are not technically complex, considerable experience and expertise may be required to design and implement a successful bioremediation programme, due to the need to thoroughly assess a site for suitability and to optimise conditions to achieve a satisfactory result.

Because bioremediation seems to be a good alternative to conventional clean-up technologies research in this field, especially in the United States, rapidly increasing. Bioremediation has been used at a number of sites worldwide, including Europe, with varying degrees of success. Techniques are improving as greater knowledge and experience are gained, and there is no doubt that bioremediation has great potential for dealing with certain types of site contamination. Unfortunately, the principles, techniques, advantages, and disadvantages of bioremediation are not widely known or understood, especially among those who will have to deal directly with bioremediation proposals, such as site owners and regulators. Here, we intended to assist by providing a straightforward, pragmatic view of the processes involved in bioremediation, the pros and cons of the technique, and the issues to be considered when dealing with a proposal for bioremediation. Some tests make an exhaustive examination of the literature of bioremediation of organic and inorganic pollutants, and another test takes a look at pertinent field application case histories.

Principles of Bioremediation

Environmental biotechnology is not a new field; composting and wastewater treatments are familiar examples of old environmental biotechnologies. However, recent studies in molecular biology and ecology offer opportunities for more efficient biological processes. Notable accomplishments of these studies include the clean-up of polluted water and land areas.

Bioremediation is defined as the process whereby organic wastes are biologically degraded under controlled conditions to an innocuous state, or to levels below concentration limits established by regulatory authorities.

By definition, bioremediation is the use of living organisms, primarily microorganisms, to degrade the environmental contaminants into less toxic forms. It uses naturally occurring bacteria and fungi or plants to degrade or detoxify substances hazardous to human health and/or the environment. The microorganisms may be indigenous to a contaminated area or they may be isolated from elsewhere and brought to the contaminated site. Contaminant compounds are transformed by living organisms through reactions that take place as a part of their metabolic processes. Biodegradation of a compound is often a result of the actions of multiple organisms. When microorganisms are imported to a contaminated site to enhance degradation we have a process known as bioaugmentation.

For bioremediation to be effective, microorganisms must enzymatically attack the pollutants and convert them to harmless products. As bioremediation can be effective only where environmental conditions permit microbial growth and activity, its application often involves the manipulation of environmental parameters to allow microbial growth and degradation to proceed at a faster rate.

Like other technologies, bioremediation has its limitations. Some contaminants, such as chlorinated organic or high aromatic hydrocarbons, are resistant to microbial attack. They are degraded either slowly or not at all, hence it is not easy to predict the rates of clean-up for a bioremediation exercise; there are no rules to predict if a contaminant can be degraded. Bioremediation techniques are typically more economical than traditional methods such as incineration, and some pollutants can be treated on site, thus reducing exposure risks for clean-up personnel, or potentially wider exposure as a result of transportation accidents. Since bioremediation is based on natural attenuation the public considers it more acceptable than other technologies. Most bioremediation systems are run under aerobic conditions, but running a system under anaerobic conditions may permit microbial organisms to degrade otherwise recalcitrant molecules.

Factors of Bioremediations

The control and optimisation of bioremediation processes is a complex system of many factors. These factors include: the existence of a microbial population capable of degrading the pollutants; the availability of contaminants to the microbial population; the environment factors (type of soil, temperature, pH, the presence of oxygen or other electron acceptors, and nutrients).

Microbial Populations for Bioremediation Processes

Microorganisms can be isolated from almost any environmental conditions. Microbes will adapt and grow at subzero temperatures, as well as extreme heat, desert conditions, in water, with an excess of oxygen, and in anaerobic conditions, with the presence of hazardous compounds or on any waste stream. The main requirements are an energy source and a carbon source. Because of the adaptability of microbes and other biological systems, these can be used to degrade or remediate environmental hazards.

We can subdivide these microorganisms into the following groups:

Aerobic (in the Presence of Oxygen): Examples of aerobic bacteria recognised for their degradative abilities are *Pseudomonas*, *Alcaligenes*, *Sphingomonas*, *Rhodococcus*, and *Mycobacterium*. These microbes have often been reported to degrade pesticides and hydrocarbons, both alkanes and polyaromatic compounds. Many of these bacteria use the contaminant as the sole source of carbon and energy.

Anaerobic (in the Absence of Oxygen): Anaerobic bacteria are not as frequently used as aerobic bacteria. There is an increasing interest in anaerobic bacteria used for bioremediation of polychlorinated biphenyls (PCBs) in river sediments, dechlorination of the solvent trichloroethylene (TCE), and chloroform.

Ligninolytic Fungi: Fungi such as the white rot fungus *Phanaerochaete chrysosporium* have the ability to degrade an extremely diverse range of persistent or toxic environmental pollutants. Common substrates used include straw, saw dust, or corn cobs.

Methylotrophs: Aerobic bacteria that grow utilising methane for carbon and energy. The initial enzyme in the pathway for aerobic degradation, methane monooxygenase, has a broad substrate range and is active against a wide range of compounds, including the chlorinated aliphatics trichloroethylene and 1,2-dichloroethane.

For degradation it is necessary that bacteria and the contaminants be in contact. This is not easily achieved, as neither the microbes nor contaminants are uniformly spread in the soil. Some bacteria are mobile and exhibit a chemotactic response, sensing the contaminant and moving towards it. Other microbes such as fungi grow in a filamentous form towards the contaminant. It is possible to enhance the mobilisation of the contaminant utilising some surfactants such as sodium dodecyl sulphate (SDS)

Environmental Factors

Nutrients: Although the microorganisms are present in contaminated soil, they cannot necessarily be there in the numbers required for bioremediation of the site. Their growth and activity must be stimulated. Biostimulation usually involves the addition of nutrients and oxygen to help indigenous microorganisms. These nutrients are the basic building blocks of life and allow microbes to create the necessary enzymes to break down the contaminants. All of them will need nitrogen, phosphorous, and carbon.

Carbon is the most basic element of living forms and is needed in greater quantities than other elements. In addition to hydrogen, oxygen, and nitrogen it constitutes about 95 per cent of the weight of cells. Phosphorous and sulphur contribute with 70 per cent of the remainders. The nutritional requirement of carbon to nitrogen ratio is 10:1, and carbon to phosphorous is 30:1.

Environmental Requirements

Microbial growth and activity are readily affected by pH, temperature, and moisture. Although microorganisms have been also isolated in extreme conditions, most of them grow optimally over a narrow range, so that it is important to achieve optimal conditions. If the soil has too much acid it is possible to rinse the pH by adding lime. Temperature affects biochemical reactions rates, and the rates of many of them double for each 10°C rise in temperature.

Above a certain temperature, however, the cells die. Plastic covering can be used to enhance solar warming in late spring, summer, and autumn. Available water is essential for all the living organisms, and irrigation is needed to achieve the optimal moisture level.

The amount of available oxygen will determine whether the system is aerobic or anaerobic. Hydrocarbons are readily degraded under aerobic conditions, whereas chlorate compounds are degraded only in anaerobic ones. To increase the oxygen amount in the soil it is possible to till or sparge air. In some cases, hydrogen peroxide or magnesium peroxide can be introduced in the environment.

Soil structure controls the effective delivery of air, water, and nutrients. To improve soil structure, materials such as gypsum or organic matter can be applied. Low soil permeability can impede movement of water, nutrients, and oxygen; hence, soils with low permeability may not be appropriate for *in situ* clean-up techniques.

Bioremediation Strategies

Different techniques are employed depending on the degree of saturation and aeration of an area. *In situ* techniques are defined as those that are applied to soil and groundwater at the site with minimal disturbance.

Ex situ techniques are those that are applied to soil and groundwater at the site which has been removed from the site via excavation (soil) or pumping (water). *Bioaugmentation* techniques involve the addition of microorganisms with the ability to degrade pollutants.

***In Situ* Bioremediation**

These techniques are generally the most desirable options due to lower cost and less disturbance since they provide the treatment in place avoiding excavation and transport of contaminants. *In situ* treatment is limited by the depth of the soil that can be effectively treated. In many soils effective oxygen diffusion for desirable rates of bioremediation extend to a range of only a few centimetres to about 30 cm into the soil, although depths of 60 cm and greater have been effectively treated in some cases the most important land treatments.

Bioventing: It is the most common *in situ* treatment and involves supplying air and nutrients through wells to contaminated soil to stimulate the indigenous bacteria. Bioventing employs low air flow rates and provides only the amount of oxygen necessary for the biodegradation while minimising volatilisation and release of contaminants to the atmosphere. It works for simple hydrocarbons and can be used where the contamination is deep under the surface.

In situ biodegradation involves supplying oxygen and nutrients by circulating aqueous solutions through contaminated soils to stimulate naturally occurring bacteria to degrade organic contaminants.

It can be used for soil and groundwater. Generally, this technique includes conditions such as the infiltration of water-containing nutrients and oxygen or other electron acceptors for groundwater treatment.

Biosparging: Biosparging involves the injection of air under pressure below the water table to increase groundwater oxygen concentrations and enhance the rate of biological degradation of contaminants by naturally occurring bacteria. Biosparging increases the mixing in the saturated zone and thereby increases the contact between soil and groundwater. The ease and low cost of installing small-diameter air injection points allows considerable flexibility in the design and construction of the system.

Bioaugmentation: Bioremediation frequently involves the addition of microorganisms indigenous or exogenous to the contaminated sites. Two factors limit the use of added microbial cultures in a land treatment unit: (1) non-indigenous cultures rarely compete well enough with an indigenous population to develop and sustain useful population levels, and (2) most soils with long-term exposure to biodegradable waste have indigenous microorganisms that are effective degrades if the land treatment unit is well managed.

Ex Situ Bioremediation

These techniques involve the excavation or removal of contaminated soil from ground.

Landfarming: It is a simple technique in which contaminated soil is excavated and spread over a prepared bed and periodically tilled until pollutants are degraded. The goal is to stimulate indigenous biodegradative microorganisms and facilitate their aerobic degradation of contaminants. In general, the practice is limited to the treatment of superficial 10–35 cm of soil. Since landfarming has the potential to reduce monitoring and maintenance costs, as well as clean-up liabilities, it has received much attention as a disposal alternative.

Composting: It is a technique that involves combining contaminated soil with non-hazardous organic amendants such as manure or agricultural wastes. The presence of these organic materials supports the development of a rich microbial population and elevated temperature characteristic of composting.

Biopiles: These are a hybrid of landfarming and composting. Essentially, engineered cells are constructed as aerated composted piles. Typically used for treatment of surface contamination with petroleum hydrocarbons they are a refined version of landfarming that tend to control physical losses of the contaminants by leaching and volatilisation. Biopiles provide a favourable environment for indigenous aerobic and anaerobic microorganisms.

Bioreactors: Slurry reactors or aqueous reactors are used for *ex situ* treatment of contaminated soil and water pumped up from a contaminated plume. Bioremediation in reactors involves the processing of contaminated solid material (soil, sediment, sludge) or water through an engineered containment system. A slurry bioreactor may be defined as a containment vessel and apparatus used to create a three-phase (solid, liquid, and gas) mixing condition to increase the bioremediation rate of soilbound and water-soluble pollutants as a water slurry of the contaminated soil and biomass (usually indigenous microorganisms) capable of degrading target contaminants. In general, the rate and extent of biodegradation are greater in a bioreactor system than *in situ* or in solid-phase systems because the contained environment is more manageable and hence more controllable and predictable. Despite the advantages of reactor systems, there are some disadvantages. The contaminated soil requires pre treatment (e.g., excavation) or alternatively the contaminant can be stripped from the soil via soil washing or physical extraction (e.g., vacuum extraction) before being placed in a bioreactor.

Advantages of Bioremediation

- Bioremediation is a natural process and is, therefore, perceived by the public as an acceptable waste treatment process for contaminated material such as soil. Microbes able to degrade the contaminant increase in numbers when the contaminant is present; when the contaminant is degraded, the biodegradative population declines. The residues for the treatment are usually harmless products and include carbon dioxide, water, and cell biomass.
- Theoretically, bioremediation is useful for the complete destruction of a wide variety of contaminants. Many compounds that are legally considered to be hazardous can be transformed to harmless products. This eliminates the chance of future liability associated with treatment and disposal of contaminated material.

- Instead of transferring contaminants from one environmental medium to another, for example, from land to water or air, the complete destruction of target pollutants is possible.
- Bioremediation can often be carried out on site, often without causing a major disruption of normal activities. This also eliminates the need to transport quantities of waste off site and the potential threats to human health and the environment that can arise during transportation.
- Bioremediation can prove less expensive than other technologies that are used for clean-up of hazardous waste.

Disadvantages of Bioremediation

- Bioremediation is limited to those compounds that are biodegradable. Not all compounds are susceptible to rapid and complete degradation.
- There are some concerns that the products of biodegradation may be more persistent or toxic than the parent compound.
- Biological processes are often highly specific. Important site factors required for success include the presence of metabolically capable microbial populations, suitable environmental growth conditions, and appropriate levels of nutrients and contaminants.
- It is difficult to extrapolate from bench and pilot-scale studies to full-scale field operations.
- Research is needed to develop and engineer bioremediation technologies that are appropriate for sites with complex mixtures of contaminants that are not evenly dispersed in the environment.
- Bioremediation often takes longer than other treatment options, such as excavation and removal of soil or incineration.

- Regulatory uncertainty remains regarding acceptable performance criteria for bioremediation. There is no accepted definition of "clean", evaluating performance of bioremediation is difficult, and there are no acceptable endpoints for bioremediation treatments.

Phytoremediation

Although the application of microbe biotechnology has been successful with petroleum-based constituents, microbial digestion has met limited success for widespread residual organic and metals pollutants. Vegetation-based remediation shows potential for accumulating, immobilising, and transforming a low level of persistent contaminants. In natural ecosystems, plants act as filters and metabolise substances generated by nature. Phytoremediation is an emerging technology that uses plants to remove contaminants from soil and water. The term "phytoremediation" is relatively new, coined in 1991. Its potential for encouraging the biodegradation of organic contaminants requires further research, although it may be a promising area for the future.

We can find five types of phytoremediation techniques, classified based on the contaminant fate: phytoextraction, phytotransformation, phytostabilisation, phytodegradation, rhizofiltration, even if a combination of these can be found in nature.

Phytoextraction or Phytoaccumulation: It is the process used by the plants to accumulate contaminants into the roots and aboveground shoots or leaves. This technique saves tremendous remediation cost by accumulating low levels of contaminants from a widespread area. Unlike the degradation mechanisms, this process produces a mass of plants and contaminants (usually metals) that can be transported for disposal or recycling.

Phytotransformation or Phytodegradation: It refers to the uptake of organic contaminants from soil, sediments, or water and, subsequently, their transformation to more stable, less toxic, or less mobile form. Metal chromium can be reduced from hexavalent to trivalent chromium, which is a less mobile and non-carcinogenic form.

Phytostabilisation: It is a technique in which plants reduce the mobility and migration of contaminated soil. Leachable constituents are adsorbed and bound into the plant structure so that they form a stable mass of plant from which the contaminants will not re-enter the environment.

Phytodegradation or Rhizodegradation: It is the breakdown of contaminants through the activity existing in the rhizosphere. This activity is due to the presence of proteins and enzymes produced by the plants or by soil organisms such as bacteria, yeast, and fungi. Rhizodegradation is a symbiotic relationship that has evolved between plants and microbes. Plants provide nutrients necessary for the microbes to thrive, while microbes provide a healthier soil environment.

Rhizofiltration: It is a water remediation technique that involves the uptake of contaminants by plant roots. Rhizofiltration is used to reduce contamination in natural wetlands and estuary areas.

Phytoremediation is well suited for use at very large field sites where other methods of remediation are not cost effective or practicable; at sites with a low concentration of contaminants where only polish treatment is required over long periods of time; and in conjunction with other technologies where vegetation is used as a final cap and closure of the site. There are some limitations to the technology that it is necessary to consider carefully before it is selected for site remediation: long duration of time for remediation, potential contamination of the vegetation and food chain, and difficulty establishing and maintaining vegetation at some sites with high toxic levels.

Genetic Engineering Approaches

The use of genetic engineering to create organisms specifically designed for bioremediation has great potential. The bacterium *Deinococcus radiodurans* (the most radioresistant organism known) has been modified to consume and digest toluene and ionic mercury from highly radioactive nuclear waste.

Advantages

There are a number of cost/efficiency advantages to bioremediation, which can be employed in areas that are inaccessible without excavation. For example, hydrocarbon spills (specifically, petrol spills) or certain chlorinated solvents may contaminate groundwater, and introducing the appropriate electron acceptor or electron donor amendment, as appropriate, may significantly reduce contaminant concentrations after a lag time allowing for acclimation. This is typically much less expensive than excavation followed by disposal elsewhere, incineration or other *ex situ* treatment strategies, and reduces or eliminates the need for "pump and treat", a common practice at sites where hydrocarbons have contaminated groundwater.

Monitoring Bioremediation

The process of bioremediation can be monitored indirectly by measuring the *Oxidation Reduction Potential* or redox in soil and groundwater, together with pH, temperature, oxygen content, electron acceptor/donor concentrations, and concentration of breakdown products (e.g. carbon dioxide). This table shows the (decreasing) biological breakdown rate as function of the redox potential.

Process	*Reaction*	*Redox potential (E_h in mV)*
aerobic:	$O_2 + 4e + 4H^+ \longrightarrow 2H_2O$	600 — 400
anaerobic:		
denitrification	$2NO_3^- + 10e^- + 12H^+ \longrightarrow N_2 + 6H_2O$	500 — 200
manganese IV reduction	$MnO_2 + 2e^- + 4H^+ \longrightarrow Mn_2^+ + 2H_2O$	400 — 200
iron III reduction	$Fe(OH)_3 + e^- + 3H^+ \longrightarrow Fe^{2+} + 3H_2O$	300 — 100
sulphate reduction	$SO_4 2^- + 8e^- + 10H^+ \longrightarrow H_2S + 4H_2O$	0 — –150
fermentation	$2CH_2O \longrightarrow CO_2 + CH_4$	–150 — –220

This, by itself and at a single site, gives little information about the process of remediation.

1. It is necessary to sample enough points on and around the contaminated site to be able to determine contours of equal redox potential. Contouring is usually done using specialised software, e.g. using Kriging interpolation.
2. If all the measurements of redox potential show is that electron acceptors have been used up, it's in effect an indicator for total microbial activity. Chemical analysis is also required to determine when the levels of contaminants and their breakdown products have been reduced to below regulatory limits.

Desulphatibacillum Alkenivorans AK-01

Strain AK-01 was isolated from the Arthur Kill, NJ/NY waterway. This site has a history of contamination from petrochemical industry. AK-01 is a delta-proteobacterium capable of utilising C_{13}-C_{18} alkanes as growth substrates (So *et al.*, 1999). Analysis of labelled and fully deuterated metabolites shows that AK-01 activates alkanes via fumarate addition to the subterminal carbon.

Background: The ubiquitous distribution of petroleum hydrocarbons in the environment is the consequence of diagenetic processes that occur in sedimentary rock formations containing large amounts of organic matter. Heat and pressure lead to the formation of a wide variety of hydrocarbons, including alkanes, alkenes, and cyclic/polycyclic aromatic hydrocarbons (PAHs), which can seep into aquatic environments. The environmental recalcitrance of many of these compounds is governed by their high bond dissociation energies. Alkanes are the least reactive class of hydrocarbons due to their apolar sigma bonds. In the absence of high temperatures, high pressures, metal catalysts or UV light, biotransformation plays the dominant role in environmental alkane degradation. The mechanisms and genetics of aerobic hydrocarbon degradation have been described extensively. The key feature of aerobic degradation is the role of dioxygen. Oxygen is not only a physiological requirement, but serves as a reactant in the hydroxylation of both aliphatic and aromatic hydrocarbons via monooxygenase and dioxygenase enzymes.

Oxygen's key role as a reactant during aerobic hydrocarbon degradation led to the belief for many years that n-alkanes and other hydrocarbons were recalcitrant under anoxic conditions. Research during the last 15 years, however, has shown that microorganisms have adapted and evolved to utilise hydrocarbon substrates under varying reducing conditions and that they employ novel biochemical processes to activate these compounds in the absence of oxygen. These processes have been observed in marine sediments, anaerobic sludge and contaminated aquifers.

The genome of *Desulphatibacillum alkenivorans* is currently being sequenced by the Joint Genome Institute. Sequencing is paid for by a Department of Energy grant to Boris Wawrik *et al.* at Rutgers University.

Desulphococcus Oleovorans Strain Hxd3

Desulphococcus oleovorans Strain Hxd3 was isolated from the saline water phase of an oil-water separator from a northern German oil field (Aeckersberg *et al.* 1991, 1998). Hxd3 is a delta-proteobacterium capable of utilising C_{12}-C_{20} alkanes as growth substrates (Aeckersberg *et al.* 1991, 1998). Hxd3 activates alkanes via carboxylation at C_3, with subsequent elimination of the terminal and subterminal carbons, yielding a fatty acid that is one carbon shorter than the parent alkane (So *et al.* 2003). Hxd3 is the only pure culture that is known to carboxylate aliphatic hydrocarbons.

The genome of *Desulphococcus oleovorans* is currently being sequenced by the Joint Genome Institute. Sequencing is paid for by a Department of Energy grant to Boris Wawrik *et al.* at Rutgers University.

Mycofiltration

Mycofiltration is the process of using mushroom mycelium mats as biological filters. The term was coined by mycologist Paul Stamets.

Stamets originally came up with the technique to control *E. coli* in the water outflow from his property. After planting a mushroom bed in the gulch where the water was leaving, within a year the coliform count had decreased to nearly undetectable levels. He discovered that the mushroom produced crystalline entities advancing in front of the growing mycelium, disintegrating when they encountered *E. coli*. As they did so, a chemical signal was sent back to the mycelium that, in turn, generated what appeared to be a customised macrocrystal which attracted the motile bacteria by the thousands, summarily stunning them. The advancing mycelium then consumed the *E. coli*, effectively eliminating them from the environment.

Another mushroom, *Polyporus umbellata*, has been demonstrated to completely inhibit the parasite that causes malaria.

One industrial application of mycofiltration has been to prevent erosion due to water run-off. Its primary application has been on abandoned logging roads. The approach here has been to place bark and wood chips onto logging roads, and inoculate this wood debris with mycelia of native fungal species. As the wood chips decompose, the mycelial networks develop and they act as filters to prevent silt-flow. In the process, they also renew topsoils, spurring the growth of native flora and fauna.

Mycoremediation is a similar process, using mushrooms to remove contaminants from polluted environments.

Mycoremediation

Mycoremediation is a form of bioremediation, the process of using mushrooms to return an environment (usually soil) contaminated by pollutants to a less contaminated state. The term *mycoremediation* was coined by Paul Stamets and refers specifically to the use of fungal mycelia in bioremediation.

One of the primary roles of fungi in the ecosystem is decomposition, which is performed by the mycelium. The mycelium secretes extracellular enzymes and acids that break down lignin and cellulose, the two main building blocks of plant fibre. These are organic compounds composed of long chains of carbon and hydrogen, structurally similar to many organic pollutants. The key to mycoremediation is determining the right fungal species to target a specific pollutant. Certain strains have been reported to successfully degrade the nerve gases VX and sarin.

In an experiment conducted in conjunction with Battelle, a major contributor in the bioremediation industry, a plot of soil contaminated with diesel oil was inoculated with mycelia of oyster mushrooms; traditional bioremediation techniques (bacteria) were used on control plots. After four weeks, more than 95 per cent of many of the PAH (polycyclic aromatic hydrocarbons) had been reduced to non-toxic components in the mycelial-inoculated plots. It appears that the natural microbial community participates with the fungi to break down contaminants, eventually into carbon dioxide and water. Wood-degrading fungi are particularly effective in breaking down aromatic pollutants (toxic components of petroleum), as well as chlorinated compounds (certain persistent pesticides; Battelle, 2000). Mycofiltration is a similar or same process, using fungal mycelia to filter toxic waste and microorganisms from water in soil.

Landfarming

Land Farming is a bioremediation treatment process that is performed in the upper soil zone or in biotreatment cells. Contaminated soils, sediments, or sludges are incorporated into the soil surface and periodically turned over (tilled) to aerate the mixture.

Applicability

This technique has been successfully used for years in the management and disposal of oily sludge and other petroleum refinery wastes. *In situ* systems have been used to treat near surface soil contamination for hydrocarbons and pesticides. The equipment employed in land farming is typical of that used in agricultural operations. These land farming activities cultivate and enhance microbial degradation of hazardous compounds. As a rule of thumb, the higher the molecular weight (i.e., the more rings within a polycyclic aromatic hydrocarbon), the slower the degradation rate. Also, the more chlorinated or nitrated the compound, the more difficult it is to degrade.

Limitations: Factors that may limit the applicability and effectiveness of the process include:

1. Large space requirements.
2. The conditions advantageous for biological degradation of contaminants are largely uncontrolled, which increases the length of time to complete remediation, particularly for recalcitrant compounds.
3. Inorganic contaminants are not biodegraded.
4. The potential of large amounts of particulate matter released by operations.
5. The presence of metal ions may be toxic to microbes and may leach from the contaminated soil into the ground.

Hydrocarbon compounds that have been identified as being not readily degraded by land farming include creosote, pentachlorophenol (PCP), and bunker C oil.

Biological Hydrogen Production

Biological hydrogen production is done in a bioreactor based on the production of hydrogen by algae. Algae produce hydrogen under certain conditions. In the late 1990s, it was discovered that if algae are deprived of sulphur they will switch from the production of oxygen, as in normal photosynthesis, to the production of hydrogen.

Bioreactor Design Issues

- Restriction of photosynthetic hydrogen production by accumulation of a proton gradient.
- Competitive inhibition of photosynthetic hydrogen production by carbon dioxide.
- Requirement for bicarbonate binding at photosystem II (PSII) for efficient photosynthetic activity.
- Competitive drainage of electrons by oxygen in algal hydrogen production.
- *Economic Feasibility:* The energy efficiency—the conversion of sunlight into hydrogen—has to reach 7-10 per cent (alga in its natural form achieves at most a meagre 0.1 per cent).

Attempts are in progress to solve these problems via bioengineering.

Milestones

2006: Researchers from the University of Bielefeld and the University of Queensland have genetically changed the single-cell green alga *Chlamydomonas reinhardtii* in such a way that it produces an especially large amount of hydrogen. The Stm6 can, in the long run, produce five times the volume made by the wild form of alga and up to 1.6-2.0 per cent energy efficiency.

2006: Unpublished work from the University of California at Berkeley-(The programme is done by Midwest Research Institute, the operating contractor for NREL)- may have brought the technology past the economically viable 10 per cent efficiency level. By shortening the chlorophyll stacks in the photosynthetic organelles, Anastasios Melis has "probably" passed the threshold.

Research

2006: At the University of Karlsruhe, a prototype of a bioreactor containing 500-1,000 litres of algae cultures is being developed. The reactor is to be used to prove the economic feasibility of the system in the next five years.

Economics

An algae farm the size of the state of Texas would produce enough hydrogen to supply the energy needs of the whole world. About 25,000 square kilometres would be sufficient to displace gasoline use in the US; this is less than a tenth of the area devoted to growing soya in the US.

History

In 1939, a German researcher named Hans Gaffron, while working at the University of Chicago, observed that the algae he was studying, *Chlamydomonas reinhardtii* (a green-algae), would sometimes switch from the production of oxygen to the production of hydrogen. Gaffron never discovered the cause for this change and for many years other scientists failed in their attempts at its discovery. In the late 1990s, professor Anastasios Melis a researcher at the University of California at Berkeley discovered that if the algae culture medium is deprived of sulphur it will switch from the production of oxygen (normal photosynthesis), to the production of hydrogen. He found that the enzyme responsible for this reaction is hydrogenase, but that the hydrogenase lost this function in the presence of oxygen. Melis found that depleting the amount of sulphur available to the algae interrupted its internal oxygen flow, allowing the hydrogenase an environment in which it can react, causing the algae to produce hydrogen.

Chlamydomonas moeweesi is also a good strain for the production of hydrogen.

Bioaugmentation

This is simply the introduction of a group of natural microbial strain or a genetically engineered variant so as to achieve bioremediation.

Usually the step involves studying the indigenous varieties

present in the location to determine if biostimulation is possible. If the indigenous variety do not have the metabolic capability to perform the remediation process, exogenous varieties with such sophisticated pathways are introduced.

The availability of glutaraldehyde treated recombinant *E. coli* is a boon to bioaugmentation approaches. The glutaraldehyde treatment crosslinks the cell wall, rendering the bacteria unable to reproduce, which makes treatment by these "catalytic particles" environmentally and ethically sound.

Bioaugmentation was invented and perfected by George M. Robinson. He was the assistant county petroleum engineer for Santa Maria California. During the 1960's, he spent his spare time experimenting with dirty jars and various mixes of microbes.

Phytoremediation

Phytoremediation describes the treatment of environmental problems (bioremediation) through the use of plants.

The word's etymology comes from the Greek "phyto" = plant, and Latin "remedium" = restoring balance, or remediating. Phytoremediation consists in depolluting contaminated soils, water or air with plants able to contain, degrade or eliminate metals, pesticides, solvents, explosives, crude oil and its derivatives, and various other contaminants, from the mediums that contain them. It is clean, efficient, inexpensive and non-environmentally disruptive, as opposed to processes that require excavation of soil.

Various Phytoremediation Processes

A range of processes mediated by plants are useful in treating environmental problems:

- *Phytoextraction:* Uptake and concentration of substances from the environment into the plant biomass.
- *Phytostabilisation:* Reducing the mobility of substances in the environment, for example by limiting the leaching of substances from the soil.

- *Phytotransformation:* Chemical modification of environmental substances as a direct result of plant metabolism, often resulting in their inactivation, degradation (phytodegradation) or immobilisation (phytostabilisation).
- *Phytostimulation:* Enhancement of soil microbial activity for the degradation of contaminants, typically by organisms that associate with roots. This process is also known as rhizosphere degradation.
- *Phytovolatilisation:* Removal of substances from soil or water with release into the air, sometimes as a result of phytotransformation to more volatile and/or less polluting substances.
- *Rhizofiltration:* Filtering water through a mass of roots to remove toxic substances or excess nutrients. The pollutants remain absorbed in or adsorbed by the roots.

Phytoextraction

Phytoextraction (or *phytoaccumulation*) uses plants to remove contaminants from soils, sediments or water into harvestable plant biomass. Phytoextraction has been growing rapidly in popularity worldwide for the last twenty years or so. Generally this process has been tried more often for extracting heavy metals than for organics. At the time of disposal contaminants are typically concentrated in the much smaller volume of the plant matter than in the initially contaminated soil or sediment. 'Mining with plants', or phytomining, is also being experimented with.

The plants absorb contaminants through the root system and store them in the root biomass and/or transport them up into the stems and/or leaves. A living plant may continue to absorb contaminants until it is harvested. After harvest a lower level of the contaminant will remain in the soil, so the growth/ harvest cycle must usually be repeated through several crops to achieve a significant cleanup. After the process, the cleaned soil can support other vegetation.

Two Versions of Phytoextraction:

- Natural Hyperaccumulation, where plants naturally takę up the contaminants in soil unassisted, and
- Induced or Assisted Hyperaccumulation, in which a conditioning fluid containing a chelator or another agent is added to soil to increase metal solubility or mobilisation so that the plants can absorb them more easily.

Examples of Phytoextraction from Soils:

- Arsenic, using the Sunflower (*Helianthus annuus*), or the Chinese Brake fern ("Pteris spp"), a hyperaccumulator. Chinese Brake fern stores arsenic in its leaves.
- Cadmium and zinc, using alpine pennycress (*Thlaspi caerulescens*), a hyperaccumulator of these metals at levels that would be toxic to many plants. On the other hand, the presence of copper seems to impair its growth.
- Lead, using Indian Mustard (*Brassica juncea*), Ragweed (*Ambrosia artemisiifolia*), Hemp Dogbane (*Apocynum cannabinum*), or Poplar trees, which sequester lead in its biomass.
- Salt-tolerant (moderately halophytic) barley and/or sugar beets are commonly used for the extraction of Sodium chloride (common salt) to reclaim fields that were previously flooded by sea water.
- Uranium, using sunflowers, as used after the Chernobyl accident.

Phytotransformation

In the case of organic pollutants, such as pesticides, explosives, solvents, industrial chemicals, and other xenobiotic substances, certain plants, such as Cannas, render these substances non-toxic by their metabolism. In other cases, microorganisms living in association with plant roots may metabolise these substances in soil or water.

Role of Genetics

Breeding programmes and genetic engineering are powerful methods for enhancing natural phytoremediation capabilities, or for introducing new capabilities into plants. Genes for phytoremediation may originate from a microorganism or may be transferred from one plant to another variety better adapted to the environmental conditions at the cleanup site.

Advantages and Limitations

- ***Advantages:***
  - the cost of the phytoremediation is lower than that of traditional processes both *in situ* and *ex situ;*
  - the plants can be easily monitored;
  - the possibility of the recovery and re-use of valuable metals (by companies specialising in "phytomining");
  - it is the least harmful method because it uses naturally occurring organisms and preserves the natural state of the environment.
- **Limitations**:
  - phytoremediation is limited to the surface area and depth occupied by the roots;
  - slow growth and low biomass require a long-term commitment;
  - with plant-based systems of remediation, it is not possible to completely to prevent the leaching of contaminants into the groundwater (without the complete removal of the contaminated ground which in itself does not resolve the problem of contamination);
  - the survival of the plants is affected by the toxicity of the contaminated land and the general condition of the soil;

– possible bioaccumulation of contaminants which then pass into the food chain, from primary level consumers upwards.

Hyperaccumulators and Biotic Interactions

A plant is said to be a hyperaccumulator if it can concentrate the pollutants in a minimum percentage which varies according to the pollutant involved for example: more than 1000 mg/kg of dry weight for nickel, copper, cobalt, chromium or lead; or more than 10,000 mg/kg for zinc or manganese. Most of the 215 metal-hyperaccumulating species included in their review hyperaccumulate nickel.

They listed 145 hyperaccumulators of nickel, 26 of cobalt, 24 of copper, 14 of zinc, four of Lead, and two of Chromium. This capacity for accumulation is due to *hypertolerance*, or *phytotolerance*: the result of adaptative evolution from the plants to hostile environments along multiple generations.

Boyd and Martens list 4 biotic interactions that may be affected by metal hyperaccumulation, to which can be added the biofilm as a particular aspect of micorrhizae:

1. Protection
2. Interferences with neighbour plants of different species
3. Mutualism (Mycorrhizal associations or micorrhizae, and Pollen and seed dispersal)
4. Commensalism
5. The biofilm.

Protection

More and more evidence show that the metals in hyperaccumulating plants give them some protection from various bacteria, fungi and/or insects.

For instance, with foliar Ni concentrations as low as 93 mg/kg, the larval weight of *Spodoptera exigua* (*Lepidoptera: Noctuidae*) (beet army worm) is reduced and time to pupation extended.

Information Published Supporting the Defence Hypothesis of Metal Hyperaccumulation

Paper	*Plant species*	*Metal*	*Organism(s) affected*
Ernst 1987	*Silene vulgaris* (Moench) Garke	Cu (400 mg g^{-1})	*Hadena cucubalis* Schiff. (*Lepidoptera: Noctuidae*)
Boyd *et al.* 1994	*Streptanthus polygaloides* Gray	Ni	*Xanthomonas campestris* (Gram-negative bacterium)
Boyd *et al.* 1994	*Streptanthus polygaloides* Gray	Ni	*Alternaria brassicicola* (Imperfect fungus)
Boyd *et al.* 1994	*Streptanthus polygaloides* Gray	Ni	*Erisyphe polygoni* (Powdery mildew)
Martens & Boyd 1994	*Streptanthus polygaloides*	Ni	*Pieris rapae* L. (*Lepidoptera: Pieridae*)
Boyd & Martens 1994	*Thlaspi montanum* L. var. *montanum*	Ni	*Pieris rapae*
Pollard & Baker 1997 J. and C. Presl.	*Thlaspi caerulescens*	Zn	*Schistocerca gregaria* (Forsk.) (*Orthoptera: Acrididae*)
Pollard & Baker 1997 J. and C. Presl.	*Thlaspi caerulescens*	Zn	*Deroceras carvanae* (Pollonera) (*Pulmonata: Limacidae*)
Pollard & Baker 1997 J. and C. Presl.	*Thlaspi caerulescens*	Zn	*Pieris brassicae* L. (*Lepidoptera: Pieridae*)

The defence against viruses is not always supported. Davis *et al.* (2001) have compared two close species *S. polygaloides* Gray (Ni hyperaccumulator) and *S. insignis Jepson* (non-accumulator), inoculating them with Turnip mosaic virus. They showed that the presence of nickel weakens the plant's response to the virus.

Circumvention of plants' elemental defences by their predators may occur in three ways: (1) selective feeding on low-metal tissues, (2) use of a varied diet to dilute metal-containing food (likely more efficient in large-sized herbivores), and (3) tolerance of high dietary metal content.

Avoidance of an Elemental Defence via Selective Feeding:

Mishra & Kar (1974) reported nickel to be transported through the xylem of crop plants. Similarly, Kramer *et al.* (1996) showed that Ni is transported as a complex with the amino-acid histidine in the xylem. This implies that phloem fluid may contain little nickel; thus phloem fluid may be used by able organisms as a rich source of carbohydrates.

Pea aphids (*Acyrthosiphon pisum* [Harris]; *Homoptera*: *Aphididae*) feeding on *Streptanthus polygaloides* Gray (*Brassicaceae*) have equal survival and reproduction rates for plants containing ca. 5000 mg/kg nickel amended with $NiCl_2$, and those containing 40 mg/kg nickel. This means that either the phloem fluid is poor in nickel even for nickel hyperaccumulators, or that the aphids tolerate nickel. Moreover the aphids feeding on high nickel-content plants only show a small increase of nickel content in their bodies, relatively to the nickel content of aphids feeding on low-nickel plants. On the other hand, aphids (*Brachycaudus lychnidis* L.) fed on the zinc-tolerant plant *Silene vulgaris* (Moench) Garcke (*Caryophyllaceae*)—which can contain up to 1400 mg/kg zinc in its leaves—were reported showing elevated (9000 mg/kg) zinc in their bodies.

Metal Tolerance: Hopkin (1989) and Klerks (1990) demonstrated it for animal species; Brown & Hall (1990) for fungal species; and Schlegel & al. (1992) and Stoppel & Schlegel (1995) for bacterial species.

Plants of *Streptanthus polygaloides* (*Brassicaceae*, Ni hyperaccumulator) can be parasitised by *Cuscuta californica* var. *breviflora* Engelm. (*Cuscutaceae*). Metal contents of *Cuscuta* ranged from 540–1220 mg/kg Ni, 73-fold higher than the metal contents of *Cuscuta* parasitising a co-occurring non-hyperaccumulator plant species. Cuscuta plants are, therefore, very Ni-tolerant - 10 mg Ni/kg is sufficient for growth to start decreasing in unadapted plants. According to Boyd & Martens (subm.) this is "the first well-documented instance of the transfer of elemental defences from a hyperaccumulating host to a seed plant parasite".

Interferences with Neighbour Plants of Different Species

Its likelihood between hyperaccumulators and neighbouring plants was suggested but no mechanism was proposed. Gabrielli *et al.* (1991), and Wilson & Agnew (1992), suggested a decrease in competition experienced by the hyperaccumulators for the litterfall from hyperaccumulators' canopy.

This mechanism mimics allelopathy in its effects, although technically due to redistribution of an element in the soil rather than to the plant manufacturing an organic compound. Boyd et Martens call it "elemental allelopathy"—without the autoxicity problem met in other types of allelopathy (Newman 1978).

Mutualism

Two types of mutualism are considered here, mycorrhizal associations or *mycorrhizae,* and animal-mediated pollen or seed dispersal.

Mycorrhizal Associations or Mycorrhizae: There are two types of mycorrhizal fungi: ectomycorrhizae and endomycorrhizae. Ectomycorrhizae form sheaths around plant roots, endomycorrhizae enter cortex cells in the roots.

Mycorrhizae are the symbiotic relationship between a soil-borne fungus and the roots of a plant. Some hyperaccumulators may form mycorrhizae and, in some cases, the latter may have a role in metal treatment. In soils with low metal levels, vesicular arbuscular mycorrhizae enhance metal uptake of non-hyperaccumulating species. On the other hand, some mycorrhizae increase metal tolerance by decreasing metal uptake in some low-accumulating species. Mycorrhizae thus assists *Calluna* in avoiding Cu and Zn toxicity. Most roots need about 100 times the amount of carbon than do the hyphae of its associated ectomycorrhizae in order to develop across the same amount of soil. It is, therefore, easier for hyphae to acquire elements that have a low mobility than it is for plant roots. Cesium-137 and strontium-90 both have low mobilities.

Mycorrhizal fungi depend on host plants for carbon, while enabling host plants to absorb the soil's nutrients and water with more efficiency. In mycorrhizae, nutrient uptake is enhanced for the plants while they provide energy-rich organic compounds to the fungus. Although certain plant species that are normally symbiotic with mycorrhizal fungi can exist without the fungal association, the fungus greatly enhances the plant's growth. Hosting mycorrhizae is much more energy effective to the plant than producing plant roots.

The *Brassicaceae* family reportedly forms few mycorrhizal associations. But Hopkins (1987) notes mycorrhizae associated with *Streptanthus glandulosus* Hook. (*Brassicaceae*), a non-accumulator. Some fungi tolerate easily the generally elevated metal contents of serpentine soils. Some of these fungal species are mycorrhizal. High levels of phosphate in the soil inhibit mycorrhizal growth.

The uptake of radionuclides by fungi depends on its nutritional mechanism (mycorrhizal or saprophytic). *Pleurotus eryngii* absorbs Cs best over Sr and Co, while *Hebeloma cylindrosporum* favours Co. But increasing the amount of K increases the uptake of Sr (chemical analogue to Ca) but not that of Cs (chemical analogue to K). Moreover, the uptake of Cs decreases with *Pleurotus eryngii* (mycorrhizal) and *Hebeloma cylindrosporum* (saprophytic) if the Cs content is increased, but that of Sr increases if its content is increased – this would indicate that the uptake is independent from the nutritional mechanism.

Pollen and Seed Dispersal: Some animals obtain food from the plant (nectar, pollen, or fruit pulp - Howe & Westley 1988). Animals feeding from hyperaccumulors high in metal content must either be metal-tolerant or dilute it with a mixed diet. Alternatively hyperaccumulators may rely on abiotic vectors or non-mutualistic animal vectors for pollen or seed transport, but we lack information on seed and pollen dispersal mechanisms for hyperaccumulating plants.

Jaffré & Schmid 1974; Jaffré *et al.* 1976; Reeves *et al.* 1981; have studied metal contents of entire flowers and/or fruits. They have recorded elevated metal levels in these. We find an exception with *Walsura monophylla* Elm. (*Meliaceae*), originating from the Philippines and showing 7000 mg/kg Ni in leaves but only 54 mg/kg in fruits. Some plants may thus have a mechanism by which metal or other contaminants is excluded from their reproductive structures.

Commensalism

This is an interaction benefiting one organism while being of neutral value to another. The most likely one with hyperaccumulators would be epiphytism. But this is most noticeable in humid habitats, whereas only a few detailed field studies of hyperaccumulators have been conducted in such habitats, and those studies (mostly to do with humid tropical forests on serpentine soils) pay little or no attention to that point (e.g., Proctor *et al.* 1989; Baker *et al.* 1992). Proctor *et al.* (1988) studied the tree *Shorea tenuiramulosa*, which can accumulate up to 1000 mg Ni/kg dry weight in leaf material.

They estimated covers of epiphytes on the boles of trees in Malaysia, but did not report values for individual species. Boyd *et al.* (1999) studied the occurrence of epiphytes on leaves of the Ni hyperaccumulating tropical shrub *Psychotria douarrei* (Beauvis.). Epiphyte load increased significantly with increasing leaf age, up to 62 per cent for the oldest leaves. An epiphyte sample of leafy liverworts removed from *P. douarrei*, was found to contain 400 mg Ni /kg dry weight (far less than the host plant, whose oldest and most heavily epiphytised leaves contained a mean value of 32,000 mg Ni/kg dry weight). High doses of Ni, therefore, do not prevent colonisation of *Psychotria douarrei* by epiphytes.

Chemicals that mediate host-epiphyte interactions are most likely to be located in the outermost tissues of the host (Gustafsson & Eriksson 1995). Also, most of the metal accumulates in epidermal or subepidermal cell walls or vacuoles (Ernst & Weinert 1972; Vazquez *et al.* 1994; Mesjasz-Rzybylowicz *et al.* 1996; Gabrielli *et al.* 1997). These findings suggest that epiphytes would experience higher metal levels when growing on hyperaccumulator leaves. But Severne (1974) measured the release of metal via leaching of leaves from the Ni hyperaccumulator *Hybanthus floribundus* (Lindl.) F. Muell. (*Violaceae*) from western Australia; he concluded that its leaves do not easily leach Ni.

In theory another commensal interaction could exist, if the high metal content of the soil under hyperaccumulator plants was needed for another plant species to establish itself. No evidence is known showing such effect.

Biostimulation

Biostimulation involves the modification of the environment to stimulate existing bacteria capable of bioremediation. This can be done by addition of various forms of limiting nutrients, such as phosphorus, nitrogen, or carbon (e.g. in the form of molasses). Additives are usually added to the subsurface through injection wells, although injection well technology for biostimulation purposes is still emerging. Removal of the contaminated material is also an option, albeit an expensive one. Biostimulation is an alternative to bioaugmentation.

The primary advantage of biostimulation is that bioremediation will be undertaken by already present native mocroorganisms that are well suited to the subsurface environment, and are well distributed spatially within the subsurface. The primary disadvantage is that the delivery of additives in a manner that allows the additives to be readily available to subsurface microorganisms is based on the local geology of the subsurface. Tight, impearmeable subsurface lithology (tight clays or other fine-grained material) make it difficult to spread additives throughout the affected area. Fractures in the subsurface create preferential pathways in the subsurface which additivies preferentially follow, preventing even distribution of additives.

Investigations to determine subsurface characteristics (such as natural groundwater velocity during ambient conditions, hydraulic conductivity of the subsurface, and lithology of the subsurface) are important in developing a successful biostimulation system. In addition, a pilot-scale study of the potential biostimulation system should be undertaken prior to full-scale design and implementation.

Deinococcus Radiodurans in Bioremedial Use

Deinococcus radiodurans ("strange berry that withstands radiation", formerly called *Micrococcus radiodurans*) is an extremophilic bacterium, and is the most radioresistant organism known. While a dose of 10 Gy is sufficient to kill a human, and a dose of 60 Gy is sufficient to kill all cells in a culture of *E. coli, D. radiodurans* is capable of withstanding an instantaneous dose of up to 5,000 Gy with no loss of viability, and an instantaneous dose of up to 15,000 Gy with 37 per cent viability. It can survive heat, cold, dehydration, vacuum, and acid, and because of its resistance to more than one extreme condition, *D. radiodurans* is known as a polyextremophile.

It has also been listed as the world's toughest bacterium in "The Guiness Book of World Records" because of its extraordinary resistance to several extreme conditions. Studies are being conducted to verify the origin of *D. radiodurans* as many scientists are speculating that it has originated on Mars. It has been classified as a Gram-positive bacterium.

The term *Deinobacter* has been replaced by *Deinococcus* based on evaluation of ribosomal RNA sequences. Several other species within the genus have been described, and they are related to heat-resistant bacteria such as *Thermus*; the group is accordingly known as *Deinococcus-Thermus*.

History

D. radiodurans was discovered in 1956 by A.W. Anderson at the Oregon Agricultural Experiment Station in Corvallis, Oregon. Experiments were being performed to determine if canned food could be sterilised using high doses of gamma radiation. A tin of meat was exposed to a dose of radiation that was thought to kill all known forms of life, but the meat subsequently spoiled. *D. radiodurans* was isolated from the meat.

Radioactivity Resistance Mechanisms

Deinococcus accomplishes its resistance to radiation by having multiple copies of its genome and rapid DNA repair mechanisms. It usually repairs breaks in its chromosomes within 12-24 hours through a 2-step process. First, *D. radiodurans* reconnects some chromosome fragments through a process called single-strand annealing. In the second step, a protein mends double-strand breaks through homologous recombination. As a consequence of its hardiness it has been nicknamed "Conan the Bacterium" (after Conan the Barbarian).

A persistent question regarding *D. radiodurans* is how such a high degree of radioresistance could evolve. Natural background radiation levels are very low — in most places, on the order of 0.4 mGy per year, and the highest known background radiation, near Guarapari, Brazil is only 175 mGy per year. With naturally-occurring background radiation levels so low, organisms evolving mechanisms specifically to ward off the effects of high radiation are unlikely.

Valerie Mattimore and John R. Battista of Louisiana State University have suggested that the radioresistance of *D. radiodurans* is simply a side-effect of a mechanism for dealing with prolonged cellular desiccation (dryness). To support this hypothesis, they performed an experiment in which they demonstrated that mutant strains of *D. radiodurans* which are highly susceptible to damage from ionising radiation are also highly susceptible to damage from prolonged desiccation, while the wild type strain is resistant to both. In addition to DNA repair, *D. radiodurans* use LEA (Late Embryogenesis Abundant) protein expression to protect against desiccation.

Michael Daly of the Uniformed Services University of the Health Sciences suggests that the bacterium uses manganese to protect itself against radiation damage.

Scanning electron microscopy analysis has shown that DNA in *D. radiodurans* is organised into tightly packed toroids, which may facilitate DNA repair.

A team of Croatian and French researchers have bombarded *D. radiodurans* to study the mechanism of DNA repair. At least two copies of the genome, with random DNA breaks, can form DNA fragments through annealing. Partially overlapping fragments are then used for synthesis of homologous regions through a moving D-loop that can continue extension until they find complementary partner strands. In the final step, there is crossover by means of RecA-dependent homologous recombination.

A team of Russian and American scientists proposed that the radioresistance of *D. radiodurans* had a Martian origin. Evolution of the microorganism could have taken place on the Martian surface until it was delivered to Earth on a meteorite. It has been hypothesised that such objects might have attained sufficient kinetic energy to achieve escape velocity as a result of a major meteorite impact on the Martian surface (this being part of a hypothecised process known as panspermia).

The heating effects of that impact and of entry into the Earth's atmosphere would have had to have been insufficient to sterilise such an object. However, apart from its resistance to radiation, Deinococcus is genetically and biochemically quite similar to other terrestrial life forms, arguing against an extraterrestrial origin.

Applications

Using genetic engineering *Deinococcus* has been used for bioremediation to consume and digest solvents and heavy metals, even in a highly radioactive site. The bacterial mercuric reductase gene has been cloned from *Escherichia coli* into *Deinococcus* to detoxify the ionic mercury frequently found in radioactive waste generated from nuclear weapons manufacture. Those researchers developed a strain of

Deinococcus that could detoxify both mercury and toluene in mixed radioactive wastes.

Some have speculated that mechanisms of DNA repair used by *D. radiodurans* could be incorporated into the genome of higher species as a means of rejuvenation.

Some scientists have suggested that *D. radiodurans* could be genetically manipulated to produce various medicines.

The Craig Venter Institute has used a system derived from the rapid DNA repair mechanisms of *D. radiodurans* to assemble synthetic genes into chromosomes to make synthetic organisms.

Bioremediation of Polluted Environment

The population explosion in the world has resulted in an increase in the area of polluted soil and water. As the number of people continues increasing day by day it also brings with it a growing pressure on air, water and land resources. In order to cater to the demands of the people, the rapid expansion of industries, food, health care, vehicles, etc. is necessary. But it is very difficult to maintain the quality of life with all these new developments, which are unfavourable to the environment in which we live, if proper management is not applied. In nature there are various fungi, bacteria and microorganisms that are constantly at work to break down organic compounds but the question arises when pollution occurs, who will do this clean up job? Since the quality of life is inextricably linked to the overall quality of the environment, global attention has been focused on ways to sustain and preserve the environment. This endeavour is possible by involving biotechnology.

Role of Environmental Biotechnology in Pollution Management

Biotechnology can be used to assess the well being of ecosystems, transform pollutants into benign substances, generate biodegradable materials from renewable sources, and develop environmentally safe manufacturing and disposal processes.

Environmental biotechnology employs the application of genetic engineering to improve the efficiency and cost, which are key factors in the future widespread exploitation of microorganisms to reduce the environmental burden of toxic substances.

In view of the urgent need of an efficient environmental biotechnological process, researchers have devised a technique called bioremediation, which is an emerging approach to rehabilitating areas fouled by pollutants or otherwise damaged through ecosystem mismanagement.

"Remediate" means to solve a problem, and "bioremediate" means to use biological organisms to solve an environmental problem such as contaminated soil or groundwater.

Bioremediation is the use of living microorganisms to degrade environmental pollutants or to prevent pollution. In other words it is a technology for removing pollutants from the environment thus restoring the original natural surroundings and preventing further pollution.

The rapid expansion and increasing sophistication of the chemical industries in the last century has meant that there has been increasing levels of complex toxic effluents being released into the environment. Many major incidents have occurred in the past which reveal the necessity to prevent the escape of effluents into the environment, such as the Exxon Valdez oil spill, the Union-Carbide (Dow) Bhopal disaster, large-scale contamination of the Rhine River, the progressive deterioration of the aquatic habitats and conifer forests in the northeastern US, Canada, and parts of Europe, or the release of radioactive material in the Chernobyl accident, etc.

Microorganisms Involved in Bioremediation Process

Many different types of organisms such as plants can be used for bioremediation but microorganisms show the greatest potential.

Microorganisms (primarily bacteria and fungi) are nature's original recyclers.

Their capability to transform natural and synthetic chemicals into sources of energy and raw materials for their own growth suggests that expensive chemical or physical remediation processes might be replaced with biological processes that are lower in cost and more environmentally friendly.

Microorganisms, therefore, represent a promising, largely untapped resource for new environmental biotechnologies. Research continues to verify the bioremediation potential of microorganisms. For example, a recent addition to the growing list of bacteria that can reduce metals is Geobacter metallireducens, which removes uranium, a radioactive waste, from drainage waters in mining operations and from contaminated groundwaters. Even dead microbial cells can be useful in bioremediation technologies.

These discoveries suggest that further exploration of microbial diversity is likely to lead to the discovery of many more organisms with unique properties useful in bioremediation.

The use of microorganisms is not limited to one field of study of bioremediation, it has an extensive use; Petroleum, its products and oils constitute hydrocarbons and if present in the environment causes pollution. Oil slicks caused by oil tankers and petrol leakage into the marine environment is now a constantly occurring phenomenon. A number of microorganisms can utilise oil as a source of food, and many of them produce potent surface-active compounds that can emulsify oil in water and facilitate its removal.

Unlike chemical surfactants, the microbial emulsifier is non-toxic and biodegradable.

The microorganisms capable of degrading petroleum include pseudomonads, various corynebacteria, mycobacteria and some yeasts.

Apart from degrading hydrocarbons, microbes also have

the ability to remove industrial wastes, reduce the toxic cations of heavy metals (such as Selenium) to a much less toxic soluble form. For example, plants like locoweed remove large amounts of the toxic element selenium. The selenium is stored in plant tissues where it poses no harm until and unless the plant is eaten. Many algae and bacteria produce secretions that attract metals that are toxic in high levels. The metals are in effect removed from the food chain by being bound to the secretions. Degradation of dyes are also brought about by some anaerobic bacteria and fungi.

To boost the world's food production rate to compensate for the increasing population, pesticides are being used. The extensive use of these artificial boosters have lead to the accumulation of artifical complex compounds called *xenobiotics*. By introducing genetically altered microbes, it is possible to degrade these compounds.

Disadvantages

Bioremediation, although considered a boon in the midst of present day environmental situations, can also be considered problematic because, while additives are added to enhance the functioning of one particular bacteria, fungi or any other microorganisms, it may be disruptive to other organisms inhabiting that same environment when done *in situ*. Even if genetically modified microorganisms are released into the environment after a certain point of time it becomes difficult to remove them. Bioremediation is generally very costly, is labour intensive, and can take several months for the remediation to achieve acceptable levels. Another problem regarding the use of *in situ* and *ex situ* processes is that it is capable of causing far more damage than the actual pollution itself.

Technologies Using Bioremediation Treatments

Technologies using bioremediation treatment include bioaugmentation, biofilters, bioreactors, biostimulation, bioventing, composting and landfarming. Despite its shortcomings, its pertinence in this world is unquestionable in the light of present day environmental hazards.

Bioremediation provides a technique for cleaning up pollution by enhancing the same biodegradation processes that occur in nature. So by developing an understanding of microbial communities and their response to the natural environment and pollutants, expanding the knowledge of the genetics of the microbes to increase capabilities to degrade pollutants, conducting field studies of new bioremediation techniques which are cost effective, and dedicating sites which are set aside for long-term research purpose, these opportunities offer potential for significant advances. There is no doubt that bioremediation is in the process of paving a way to greener pastures.

11

Molecular Cloning

Molecular cloning refers to the procedure of isolating a defined DNA sequence and obtaining multiple copies of it *in vivo*. Cloning is frequently employed to amplify DNA fragments containing genes, but it can be used to amplify any DNA sequence such as promoters, non-coding sequences and randomly fragmented DNA. It is utilised in a wide array of biological experiments and practical applications such as large scale protein production. Occasionally, the term cloning is misleadingly used to refer to the identification of the chromosomal location of a gene associated with a particular phenotype of interest, such as in positional cloning. In practice, localisation of the gene to a chromosome or genomic region does not necessarily enable one to isolate or amplify the relevant genomic sequence.

In essence, in order to amplify any DNA sequence in a living organism that sequence must be linked to an origin of replication, a sequence element capable of directing the propagation of itself and any linked sequence. In practice, however, a number of other features are desired and a variety of specialised cloning vectors exist that allow protein expression, tagging, single stranded RNA and DNA production and a host of other manipulations.

Cloning of any DNA fragment essentially involves four steps: fragmentation, ligation, transfection, and screening/ selection. Although these steps are invariable among cloning procedures a number of alternative routes can be selected, these are summarised as a 'cloning strategy'.

Initially, the DNA of interest needs to be isolated to provide a relevant DNA segment of suitable size. Subsequently, a ligation procedure is employed whereby the amplified fragment is inserted into a vector. The vector (which is frequently circular) is linearised by means of restriction enzymes, and incubated with the fragment of interest under appropriate conditions with an enzyme called DNA ligase. Following ligation the vector with the insert of interest is transfected into cells. A number of alternative techniques are available, such as chemical sensitivation of cells, electroporation and biolistics. Finally, the transfected cells are cultured. As the aforementioned procedures are of particularly low efficiency, there is a need to identify the cells that have been successfully transfected with the vector construct containing the desired insertion sequence in the required orientation. Modern cloning vectors include selectable antibiotic resistance markers, which allow only cells in which the vector has been transfected, to grow. Additionally, the cloning vectors may contain colour selection markers which provide blue/white screening (a-factor complementation) on X-gal medium. Nevertheless, these selection steps do not absolutely guarantee that the DNA insert is present in the cells obtained. Further investigation of the resulting colonies is required to confirm that cloning was successful. This may be accomplished by means of PCR, restriction fragment analysis and/or DNA sequencing.

Cellular Cloning

Cloning a cell means to derive a population of cells from a single cell. In the case of unicellular organisms such as bacteria and yeast, this process is remarkably simple and essentially only requires the inoculation of the appropriate medium. However, in the case of cell cultures from higher organisms, cell cloning is an arduous task as these cells will not readily grow in standard media.

A useful tissue culture technique used to clone distinct lineages of cell lines involves the use of cloning rings (cylinders). According to this technique, a single-cell suspension of cells which have been exposed to a mutagenic agent or drug used to drive selection is plated at high dilution to create isolated colonies; each arising from a single and potentially clonally distinct cell. At an early growth stage when colonies consist of only a few of cells, sterile polystyrene rings (cloning rings), which have been dipped in grease are placed over an individual colony and a small amount of trypsin is added. Cloned cells are collected from inside the ring and transferred to a new vessel for further growth.

Organism

Organism cloning refers to the procedure of creating a new multicellular organism, genetically identical to another. In essence this form of cloning is an asexual method of reproduction, where fertilization or inter-gamete contact does not take place. Asexual reproduction is a naturally occurring phenomenon in many species, including most plants and some insects.

Horticulture

The term *clone* is used in horticulture to mean all descendants of a single plant, produced by vegetative reproduction or apomixis. Many horticultural plant cultivars are clones, having been derived from a single individual, multiplied by some process other than sexual reproduction. As an example, some European cultivars of grapes represent clones that have been propagated for over two millennia. Other examples are potato and banana. Grafting can be regarded as cloning, since all the shoots and branches coming from the graft are genetically a clone of a single individual, although the root systems may be genetically genuine examples of cloning in the broader biological sense, as they create genetically identical organisms by biological means, but this particular kind of cloning has not come under ethical scrutiny and is generally treated as an entirely different kind of operation.

Many trees, shrubs, vines, ferns and other herbaceous perennials form clonal colonies. Parts of a large clonal colony often become detached from the parent, termed fragmentation, to form separate individuals. Some plants also form seeds asexually, termed apomixis, e.g. dandelion.

Animals

Clonal derivation exists in nature in some animal species and is referred to as parthenogenesis. An example is the "Little Fire Ant" (*Wasmannia auropunctata*), which is native to Central and South America but has spread throughout many tropical environments.

Therapeutic Cloning

Therapeutic cloning refers to a procedure which produces cells, specific body parts, and organs to be utilised for medical purposes. This has not yet been realised, but it is the subject of much active research. Currently, patients subjected to transplantation are administered immunosuppressive drugs to prevent recognition of the foreign transplant by their immune system and its subsequent rejection.

The ability to clonally derive tissues and organs from the patients' own cells would abolish the need for immunosuppressive drugs and would allow the patients to live a life without the potentially serious side-effects of immunosuppressive drugs. More importantly, the ability to clonally derive organs would alleviate the current shortage of transplants and would possibly reduce waiting times for transplants to become available.

Reproductive Cloning

Reproductive cloning uses "somatic cell nuclear transfer" (SCNT) to create animals that are genetically identical. This process entails the transfer of a nucleus from a donor adult cell (somatic cell) to an egg which has no nucleus. If the egg begins to divide normally it is transferred into the uterus of the surrogate mother.

Such clones are not strictly identical since the somatic cells may contain mutations in their nuclear DNA. Additionally, the mitochondria in the cytoplasm also contains DNA and during SCNT this DNA is wholly from the donor egg, thus the mitochondrial genome is not the same as that of the nucleus donor cell from which it was produced. This may have important implications for cross-species nuclear transfer in which nuclear-mitochondrial incompatibilities may lead to death.

Species Cloned

The modern cloning techniques involving nuclear transfer have been successfully performed on several species. Landmark experiments in chronological order:

- *Tadpole (1952):* Many scientists questioned whether cloning had actually occurred and unpublished experiments by other labs were not able to reproduce the reported results.
- *Carp (1963):* In China, embryologist Tong Dizhou cloned a fish. He published the findings in an obscure Chinese science journal which was never translated into English.
- *Sheep (1996):* From early embryonic cells by Steen Willadsen. Megan and Morag cloned from differentiated embryonic cells in June 1995 and Dolly the sheep in 1997.
- *Rhesus Monkey:* Tetra (female, January 2000) from embryo splitting.
- *Cattle:* Alpha and Beta (males, 2001) and (2005) Brazil.

- *Cat:* CopyCat "CC" (female, late 2001), Little Nicky, 2004, was the first cat cloned for commercial reasons.
- *Mule:* Idaho Gem, a john mule born 2003-05-04, was the first horse-family clone.
- *Horse:* Prometea, a Haflinger female born 2003-05-28, was the first horse clone.

Health Aspects

The success rate of cloning has been low: Dolly the sheep was born after 277 eggs were used to create 29 embryos, which only produced three lambs at birth, only one of which lived, Dolly. Seventy calves have been created from 9,000 attempts and one third of them died young; Prometea took 328 attempts, and, more recently, Paris Texas was created after 400 attempts. Notably, although the first clones were frogs, no adult cloned frog has yet been produced from a somatic adult nucleus donor cell.

There were early claims that Dolly the Sheep had accelerated ageing. Ageing of this type is thought to be due to the shortening of telomeres, regions at the tips of chromosomes which prevent genetic threads from fraying every time a cell divides. Over time telomeres get worn down until cell-division is no longer possible — this is thought to be a cause of ageing. However, subsequent studies showed that, if anything, Dolly's telomere were longer than normal. Dolly died in the year of 2003. Ian Wilmut said that Dolly's early death had nothing to do with cloning but with a respiratory infection common to lambs raised like Dolly.

Consistent with Dolly's telomeres being longer, analysis of the telomeres from cloned cows showed that they were also longer. This suggests clones could live longer life spans although many died young after excessive growth. Researchers think that this could eventually be developed to reverse ageing in humans, provided that this is based chiefly on the shortening of telomeres. Although some work has been performed on telomeres and ageing in nuclear transfer clones, the evidence is at an early stage.

Dolly the Sheep

Dolly (1996-07-05 – 2003-02-14), a Finn Dorsett ewe, was the first mammal to have been successfully cloned from an adult cell. She was cloned at the Roslin Institute in the United Kingdom and lived there until her death when she was six. On 2003-04-09 her stuffed remains were placed at Edinburgh's Royal Museum, part of the National Museums of Scotland.

Dolly's publicity was inevitable because it showed that the genetic material from a specific adult cell programmed to express only a distinct subset of its genes, could be reprogrammed to grow an entirely new organism. Before this demonstration, there was no proof for the widely spread hypothesis that differentiated animal cells can give rise to new entire new organisms.

Human Cloning

Human cloning is the creation of a genetically identical copy of an existing or previously existing human. The term is generally used to refer to *artificial* human cloning; human clones in the form of identical twins are commonplace, with their cloning occurring during the natural process of reproduction. There are two commonly discussed types of human cloning: therapeutic cloning and reproductive cloning. A third type of cloning called replacement cloning exists in theory and is a combination of therapeutic and reproductive cloning. Replacement cloning entails the replacement of an extensively damaged, failed, or failing body through cloning followed by whole or partial brain transplant. It has been proposed as a way to greatly extend lifespan.

Human cloning is among the most controversial forms of the practice. There have been numerous demands for all progress in the human cloning field to be halted. Some people and groups oppose therapeutic cloning but many more oppose reproductive cloning. The American Association for the Advancement of Science (AAAS) and other scientific organisations have made public statements suggesting that human reproductive cloning be banned until safety issues are resolved. Serious ethical issues have arisen in discussions of harvesting of organs from clones. Some people have considered the idea of growing organs separately from a human organism—in doing this, a new organ supply could be established without the moral implications of harvesting them from human organisms. Research is also being done on the idea of growing organs that are biologically acceptable to the human body inside of other organisms, such as pigs or cows, then transplanting them to humans.

Ethical Issues of Cloning

Although the practice of cloning organisms has been widespread for several thousands of years in the form of horticultural cloning, the recent technological advancements that have allowed for cloning of animals (and potentially humans) have been highly controversial. Some believe it is unethical to use a human clone to save the life of another. Others have countered that people who exist today and have interpersonal relationships and personal histories should take precedence over never-conscious life at any stage of developmental maturity.

Many religious groups oppose all forms of cloning, including the potentially life-saving cloning of individual organs, on the grounds that life begins at conception. However, the idea that life begins at conception is based on modern scientific descriptions of reproduction and discussions of conception in the texts of major religions are not factually accurate. Furthermore, in the old testament, the recommended punishment (payment of a fine decided by a judge) of a person who causes miscarriage of a human fetus clearly indicates that the fetus is not considered equivalent to a person. Concerns also exist regarding the protection of the identity of the individual and the right to protect one's genetic identity.

The social implications of an artificial human production scheme were famously explored in the novel *Brave New World* and also in the movie *The Island*. These fictional works explore the prejudices that could develop against clones with respect to a kind of slavery or caste system.

Cloning Extinct and Endangered Species

Cloning, or more precisely, the reconstruction of functional DNA from extinct species has, for decades, been a dream of some scientists. The possible implications of this were dramatised in the best-selling novel by Michael Crichton and high budget Hollywood thriller *Jurassic Park*. In real life, one of the most anticipated targets for cloning was once the Woolly Mammoth, but attempts to extract DNA from frozen mammoths have been unsuccessful, though a joint Russo-Japanese team is currently working towards this goal.

In 2000, a cow named Bessie gave birth to a cloned Asian gaur, an endangered species, but the calf died after two days. In 2003, a banteng was successfully cloned, followed by three African wildcats from a thawed frozen embryo. These successes provided hope that similar techniques (using surrogate mothers of another species) might be used to clone extinct species. Anticipating this possibility, tissue samples from the last *bucardo* (Pyrenean Ibex) were frozen immediately after it died. Researchers are also considering cloning endangered species such as the giant panda, ocelot, and cheetah. The "Frozen Zoo" at the San Diego Zoo now stores frozen tissue from the world's rarest and most endangered species.

In 2002, geneticists at the Australian Museum announced that they had replicated DNA of the Thylacine (Tasmanian Tiger), extinct about 65 years previous, using polymerase chain reaction. However, on 2005-02-15 the museum announced that it was stopping the project after tests showed the specimens' DNA had been too badly degraded by the (ethanol) preservative. Most recently, on 2005-05-15, it was announced that the Thylacine project would be revived, with new participation from researchers in New South Wales and Victoria.

One of the continuing obstacles in the attempt to clone extinct species is the need for nearly perfect DNA. Cloning from a single specimen could not create a viable breeding population in sexually reproducing animals. Furthermore, even if males and females were cloned, the question would remain open if they would be viable at all in the absence of parents that could teach or show them natural behaviour. Essentially, if cloning an extinct species succeeded — it must be considered that cloning still is an experimental technology that succeeds only by chance — it is far more likely than not that any resulting animals, even if they were healthy, would be little more than curios or museum pieces.

Cloning endangered species is a highly ideological issue. Many conservation biologists and environmentalists vehemently oppose cloning endangered species — not because they think it won't work but because they think it may deter donations to help preserve natural habitat and wild animal populations. The "rule-of-thumb" in animal conservation is that, if it is still feasible to conserve habitat and viable wild populations, breeding in captivity should not be undertaken in isolation.

In a 2006 review, David Ehrenfeld concludes that cloning in animal conservation is an experimental technology that, at its present state, cannot be expected to work except by pure chance and utterly fails a cost-benefit analysis. Furthermore, he says, it is likely to siphon funds from established and working projects and does not address any of the issues underlying animal extinction (such as habitat destruction, hunting or other overexploitation, and an impoverished gene pool). While cloning technologies are well-established and used on a regular basis in plant conservation, care must be taken to ensure genetic diversity. He concludes:

"Vertebrate cloning poses little risk to the environment, but it can consume scarce conservation resources, and its chances of success in preserving species seem poor. To date, the conservation benefits of transgenics and vertebrate cloning remain entirely theoretical, but many of the risks are known and documented. Conservation biologists should devote their research and energies to the established methods of conservation, none of which require transgenics or vertebrate cloning."

Embryo

Somatic Cell Nuclear Transfer: Somatic cell nuclear transfer can also be used to create a clonal embryo. The most likely scenario for this is to produce embryos for use in research, particularly stem cell research. This process is also called "research cloning" or "therapeutic cloning."

Therapeutic cloning, also called "embryo cloning," is the production of human embryos for use in research. The goal of this process is not to create cloned human beings, but rather to harvest stem cells that can be used to study human development and to treat disease. Stem cells are important to biomedical researchers because they can be used to generate virtually any type of specialised cell in the human body. Stem cells are extracted from the egg after it has divided for 5 days. The egg at this stage of development is called a blastocyst. The extraction process destroys the embryo, which raises a variety of ethical concerns. Many researchers hope that one day stem cells can be used to serve as replacement cells to treat heart disease, Alzheimer's, cancer, and other diseases.

Scientists believe that cloning may be used to create stem cells genetically compatible with the somatic cell donor. Cloning in stem cell research, called *research cloning* or *therapeutic cloning*, has not yet been successful: no embryonic stem cell lines have been derived from clonal embryos. The process might provide a way to grow organs in host carriers, so that organs could be produced which would be completely compatible with the original tissue donor. Host carrier growing poses a risk of transspecies diseases if the host is of a different species (e.g., a pig).

In human beings, this is a highly controversial issue for several reasons. It involves creating human embryos in vitro and then destroying them during the process of attempting to obtain embryonic stem cells. But proposals to use cloning techniques in human stem cell research raise a set of concerns beyond the moral status of the embryo. These have led a number of individuals and organisations, who are not opposed in principle to human embryonic stem cell research, to be concerned about or opposed to, human research cloning.

One concern is that cloning in human stem cell research will lead to the reproductive cloning of humans. A second concern relates to the appropriate sourcing of the eggs that are needed. Research cloning requires a large number of human eggs, which can only be obtained from women. A third concern is the feasibility of developing stem cell therapies from cloning.

In November 2001, scientists from Advanced Cell Technologies (ACT), a biotechnology company in Massachusetts, announced that they had cloned the first human embryos for the purpose of advancing therapeutic research. To do this, they collected eggs from women's ovaries and then removed the genetic material from these eggs with a needle less than 2/10,000th of an inch wide. A skin cell was inserted inside the enucleated egg to serve as a new nucleus. The egg began to divide after it was stimulated with a chemical called ionomycin. The results were limited in success. Although this process was carried out with eight eggs, only three began dividing, and only one was able to divide into six cells before stopping.

Ethics of Cloning

Roman Catholicism and many conservative Christian groups have opposed human cloning and the cloning of human embryos, believing that life begins at the moment of conception. Other Christian denominations such as the United Church of Christ do not believe a fertilized egg constitutes a living being, but still they oppose the cloning of embryonic cells. The World Council of Churches, representing nearly 400 Christian denominations worldwide, opposed cloning of both human embryos and whole humans in February 2006. The United

Methodist Church opposed research and reproductive cloning in May, 2000 and again in May, 2004.

At present, the main objection to human cloning is that the cloned individual may be biologically damaged, due to the inherent unreliability of its origin; for example, researchers currently are unable to safely and reliably clone non-human primates.

Many believe that as cloning research and methods improve, concerns of safety and reliability will no longer be an issue. However, it must be pointed out that this has yet to occur, and may never occur. Rudolph Jaenisch, a professor at Harvard, has pointed out that we have become more efficient at producing clones which are still defective.

Other arguments against cloning come from various religious orders (believing cloning violates God's will or the natural order of life), and a general discomfort some have with the idea of "meddling" with the creation and basic function of life. This unease often manifests itself in contemporary novels, movies, and popular culture, as it did with numerous prior scientific discoveries and inventions. Various fictional scenarios portray clones being unhappy, soulless, or unable to integrate into society. Furthermore, clones are often depicted not as unique individuals but as "spare parts," providing organs for the clone's original (or any non-clone that requires replacement organs).

Needless to say, cloning is a poignant and important topic, reflected by its frequent discussion and debate among politicians, scientists, the media, religions, and the general public.

On December 28, 2006, the FDA approved eating meat from cloned animals. It was said to be virtually indistinguishable from the non-cloned animals. Furthermore, companies would not be required to provide labels informing the consumer that the meat comes from a cloned animal.

CC (CAT): CC for "Carbon Copy" or "Copy Cat" (born

December 22, 2001), is a brown tabby and white domestic shorthair and the first cloned pet. CC's surrogate mother was a tabby, but her genetic donor, Rainbow, was a calico domestic shorthair. The difference in hair colouration between CC and Rainbow is due to X-inactivation. The embryo that became CC was the only one of 87 embryos produced in this research project that developed into a full-term pregnancy after being transferred to surrogate mothers.

The research project that produced CC was known as "Operation CopyCat," and was part of a larger project (called Missyplicity) to clone a dog named Missy. Entrepreneur John Sperling funded the research through his, now closed, company Genetic Savings and Clone, which provided commercial gene banking and cloning services to pet owners.

CC was born at the College of Veterinary Medicine at Texas A&M University, under the direction of Dr. Mark Westhusin, in collaboration with Dr. Taeyoung Shin. Her existence was announced publicly in February 2002 in conjunction with the publication of a scientific paper about the accomplishment.

CC now lives in the household of Dr. Duane Kraemer, one of the scientists who worked on the project.

In September 2006, CC gave birth to three kittens who were fathered naturally. This is the first time a cloned pet has given birth. CC appears to be free of health problems that seem to have arisen in other animal clones due to the cloning process. "CC has always been a perfectly normal cat and her kittens are just that way, too," says Kraemer. "We've been monitoring their health and all of them are fine, just like CC has been for the past five years."

Cumulina

Cumulina (December 3, 1997-May 5, 2000) was a mouse who was the first animal cloned from adult cells that survived to adulthood. She was cloned using the Honolulu technique developed by the Ryuzo Yanagimachi research group, 'Team

Yana', at the former campus of the John A. Burns School of Medicine located at the University of Hawai'i at Manoa. She was a brown *Mus musculus* or common house mouse. Cumulina was named after the cumulus cells surrounding the developing ovarian follicle in mice. Nuclei from these cells were put into egg cell devoid of their original nuclei in the Honolulu cloning technique. All other mice produced by the Yanagimachi lab are just known by a number.

Cumulina was able to produce two healthy litters. She was retired after the second.

One can now visit Cumulina's preserved remains at the Institute for Biogenesis Research laboratory in Honolulu, Hawaii. The Institute is a part of the John A. Burns School of Medicine.

Some of her descendants have been displayed at the Bishop Museum and the Museum of Science and Industry in Chicago, Illinois.

Idaho Gem

The mule Idaho Gem (Born May 4, 2003) was the first clone born in the horse family.

He resulted from the collaboration of Dr. Gordon Woods and Dr. Dirk Vanderwall of the Northwest Equine Reproduction Laboratory at the University of Idaho and Dr. Ken White of Utah State University. Two more healthy mule clones were born in succeeding months as a result of Project Idaho, Utah Pioneer on June 9, and Idaho Star on July 27. The project was largely financed by Post Falls, Idaho, businessman Don Jacklin, who also served as President of the American Mule Racing Association. Idaho Gem and Idaho Star were transported to trainers in 2005 to prepare them for racing in 2006. Idaho Gem and Idaho Star both won their first races on June 3, 2006, separate trial races for the Humboldt Futurity during the Winnemucca, Nev., Mule Races, Show and Draft Horse Challenge, June 3 and 4. In the June 4 futurity, Idaho Gem finished third and Idaho Star finished seventh. Idaho Gem won his next race at the San Joaquin Fair in Stockton, June 21. His time of 20.724 seconds over the 350-yard course was the fastest time by a 3-year-old mule through the end of July, the halfway point in the mule racing season. Idaho Gem also collected two seconds in photo finishes with the racing mule Out of My League. The total margin of victory between the two mules in the two races was.043 seconds. Through his first six races, Idaho Gem collected two firsts, two seconds a third and a fourth.

On June 4, 2006, Idaho Gem finished 3rd in the Winnemucca Mule Race. This was the 1st showdown between cloned and natural-born mules.

Little Nicky (Cat)

Little Nicky (born October 17, 2004) is the first commercially-produced cat clone. He was produced from the DNA of a 17-year-old cat named Nicky who died in 2003. Little Nicky's owner, a north Texas woman named Julie (her last name was not released) who paid $50,000 to have Nicky cloned, claimed that Little Nicky's behaviour is similar to that of her late cat Nicky, including an unusual liking for water.

The cloning operation was performed by the California-based company Genetic Savings and Clone, which later closed in 2006 due to lack of demand.

The Humane Society and other animal welfare groups denounced the cloning, saying that the $50,000 could have been better used to save some of the millions of pets put to sleep each year. The cat showed no negative side effects from the cloning and is alive and healthy.

Noah (Cloned Gaur)

Noah was the name of species of ox called *gaur*, cloned and gestated in the womb of a cow. Gaur is a vulnerable species according to the IUCN. The baby bull gaur was delivered on January 8, 2001, but regrettably died within just 48 hours of a common dysentery. Researchers from Advanced Cell

Technology (ACT) in Massachusetts said the problem was unlikely to be related to the cloning procedure itself.

Prometea

Prometea (born May 28, 2003) is a Haflinger foal, the first cloned horse and the first to be born from and carried by its cloning mother. Her birth was announced publicly on August 6, 2003. Born 36 kg after a natural delivery and a full-term pregnancy in Laboratory of Reproductive Technology, Cremona, Italy, Prometea at 2-month-old is 100 kg (220 pounds).

Dr. Cesare Galli and others at the lab experimented with 841 reconstructed embryos; only that of Prometea succeeded. Texas A&M University was also undertaking a horse cloning project when the Italian team first succeeded.

Horse cloning like Prometea could eliminate the problem of champion racing geldings. Jockey Club of North American thoroughbred horses has proclaimed, however, that it will allow no cloned horse in their races. The name *Prometea* is the feminine form of *Prometeos* ("Prometheus" in Italian). The horse is the seventh species to be cloned.

Bibliography

Alfoldi, L.: *Hydrogeologic Considerations for in situ Bioremediation*, Butterworth-Heinemann Publishers, Stoneham, 1991.

Anderson, T. A. and Coats, J. R.: *Bioremediation Through Rhizosphere Technology*, American Chemical Society, Washington, 1994.

Antebi, Elizabeth, and David Fishlock: *Biotechnology, Strategies for Life*, Massachusetts, Cambridge, 1986.

Bailey, Ronald: *Ecoscam, the False Prophets of Ecological Apocalypse*, St. Martin's Press, New York, 1993.

Barley, S., J. Freeman and R. Hybels: *Strategic Alliances in Commercial Biotechnology*, Harvard Business School Press, Boston, 1992.

Batra, R. and W. Klassen: *Public Perception of Biotechnology*, Agricultural Research Institute, Maryland, 1987.

Biotechnology Industry Organization (BIO): *Agriculture and the New Biology*, The Association, Washington, 1987.

Biotechnology Research Subcommittee: *Committee in Fundamental Science and Research*, National Science and Technology Council, Washington, 1995.

Blakely, E.: *The Economic Development Potentials of California's Biotech Industry, Biotech Industry Research Group, Institute of Urban and Regional Development*, University of California, Berkeley, 1989.

Blakely, Edward J. and Kelvin W. Willoughby: *Institute of Urban and Regional Development, Biotechnology Industry Research Group,* University of California, Berkeley, 1990.

Blum, Deborah: *The Monkey Wars,* Oxford University Press, London, 1994.

Booz, Allen & Hamilton: *Foreign Investment in Bay Area Bioscience,* Bay Area Bioscience Center, Oakland, 1991.

Center for Agricultural Molecular Biology: *Cook College, New Jersey Agricultural Experiment Station, Rutgers,* The State University, New Brunswick, 1991.

Chong, Ken P. and Smith, John W.: *Mechanics of Oil Shale, London, England and New York,* Elsevier Applied Science Publishers, New York, 1984.

Cirpka, O. and Bisch, G.: *Technical Design of the Aquifer Model and Influence of Spatial Variability,* Presented at the Third International Symposium on In-Situ and On-Site Bioreclamation, California, 1995.

Committee on In Situ Bioremediation Water Science and Technology Board Commission on Engineering and Technical Systems National Research Council: *In Situ Bioremediation, When does it Work,* National Academy Press, Washington, 1993.

Committee on Scientific Evaluation of the Introduction of Genetically Modified Microorganisms and Plants into the Environment: *Field Testing Genetically Modified Organisms, Framework for Decisions,* National Academy Press, Washington, 1989.

Comstock, G.: *What Obligations have Scientists to Transgenic Animals, College Station, Center for Biotechnology Policy and Ethics,* Texas A&M University, *New Jersey,* 1992.

Cooney, J. J. and Gadd, G. M.: *Metals and Microorganisms, Journal of Industrial Microbiology,* Washington, 1995.

Crawford, Ronald L. and Don L.: *Crawford, Bioremediation*

Principles and Applications, Cambridge University Press, New York, 1996.

Dale, Jeremy W.: *Molecular Genetics of Bacteria,* John Wiley and Sons, New York, 1989.

De Lorenzo, V.: *Genetic Engineering Strategies for Environmental Applications,* Current Opinion in Biotechnology, California, 1992.

Drexler, K. Nanosystems: *Molecular Machinery, Manufacturing, and Computation,* John Wiley and Sons, New York, 1992.

Economidis, I. and Palmisano A.: *Environmental Biotechnology, EC-US Task Force on Biotechnology Research,* Brussels, Belgium, 1995.

Ehrlich, H. L.: *Geomicrobiology,* Marcel Dekker, New York, 1990.

Esser, K.: *Plasmids of Eukaryotes, Fundamentals and Applications,* Springer-Verlag Berlin Heidelberg, Germany, 1986.

Fowle, John R.: *Application of Biotechnology, Environmental and Policy Issues,* Westview Press, Colorado, 1987.

Francis, A. J.: *Microbial Dissolution and Stabilization of Toxic Metals and Radionuclides in Mixed Wastes,* Oxford, New York, 1990.

Frankenberger, W. T. and Benson S. M.: *Selenium in the Environment,* Marcel Dekker, New York, 1994.

Friedman, D.: *The New Economy Project Final Report,* The New Economy Team, UCLA, 1994.

Gendel, Steven: *Agricultural Bioethics, Implications of Agricultural Biotechnology,* Iowa State University Press, Ames, 1990.

Gibson, D. T. and Sayler, G. S.: *Scientific Foundations of Bioremediation, Current Status and Future Needs,* American Academy of Microbiology, Washington, 1992.

Ginzburg, Lev. R.: *Assessing Ecological Risks of Biotechnology,* Butterworth-Heinemann, Boston, 1991.

Goodsell, David S.:*The Machinery of Life,* Springer-Verlag, New York, 1993.

Grimes, D. J.: *Validation of Reliability and Predictability of Environmental Biotechnologies,* Workshop on Bioremediation, Japan, 1995.

Grocery Manufacturers of America: *Safe Foods Developed Through Biotechnology, Nutritious, Healthful,* Abundant and Good Tasting, 1993.

Hacking, Andrew J.: *Economic Aspects of Biotechnology,* Cambridge University Press, Cambridge, 1986.

Harkness, M. R., McDermott J. B. and Abramowicz D. A.: *In Situ Stimulation of Aerobic Degradation of PCB Biodegradation in Hudson River Sediments,* MacMillan, London, 1993.

Hettinger, N.: *Owning Varieties of Life, Biotechnology, Intellectual Property, and Environmental Ethics, Center for Biotechnology Policy and Ethics,* Texas A&M University, Texas, 1994.

Hettinger, N.: *Biotechnology Policy and Ethics,* A&M University, Texas, 1992.

Hinchee, R. E., Anderson D. B., Metting, F. B. and Sayles G. D.: *Applied Biotechnology for Site Remediation,* Lewis Publishers, Florida, 1994.

Hinchee, Robert E.: *Applied Biotechnology for Site Remediation,* CRC Press, Ann Arbor, 1994.

Hollenberg, C. P. and H. Sahm: *Microbial Genetic Engineering and Enzyme Technology,* Gustav Fischer, New York, 1987.

Kelly, Kevin: *Out of Control,* Addison-Wesley, Menlo Park, 1994.

Kenney, Martin *Biotechnology: The University-Industrial Complex,* Yale University Press, Connecticut, 1986.

Kimbrell, Andrew: *The Human Body Shop,* Harper/Collins, San Francisco, 1993.

King, R. B., Mihony, G. and Sheldon J. K.: *Practical Environmental Bioremediation,* Lewis Publishers, Florida, 1992.

Kloppenburg, Jack Ralph: *The Political Economy of Plant Biotechnology,* Cambridge University Press, Cambridge, 1988.

Koehler, Gus.: *Tools to Promote Economic Success,* California Research Bureau, Sacramento, 1995.

Kotkin, Joel and Levy, Stephen: *A Twenty-First Century Prospectus, Center for the New West, and Center for the Continuing Study of the California Economy,* California, 1996.

Krimsky, Sheldon: *Biotechnics and Society,The Rise of Industrial Genetics,* Praeger, New York, 1991.

Lacy, W., L. Busch, and L. Lacy: *Agricultural Biotechnology, Issues and Choices,* Purdue University Press, Indiana, 1991.

Lacy, W., L. Lacy and L. Busch: *Emerging Trends, Consequences, and Policy Issues in Agricultural Biotechnology,* Westview Press, Colorado, 1991.

Lee, V. and Yen T.F.: *Biopolymer Shields, in Applied Biotechnology for Site Remediation,* Lewis Publishers, Florida, 1994.

Levorsen, A. I.: *Geology of Petroleum, San Francisco,* W. H. Freeman and Company, California, 1967.

Lovley, D. R.: *Dissimilatory Metal Reduction,* Annual Review of Microbiology, 1993.

Lynch, J. M. and Hobbie J. E.: *Concepts and Applications in Microbial Ecology,* Blackwell Scientific Publications, England, 1988.

McKinsey Global Institute: *Manufacturing Productivity,* McKinsey and Company, Washington, 1993.

Metting, Jr., F. B.: *Soil Microbial Ecology, Applications in Agricultural and Environmental Management,* Marcel Dekker, New York, 1992.

Miller, Henry I., M.D : *Remarks for the California Legislature*

Assembly Committee on Agriculture, MacMillan, London, 1994.

Miller, R. V. and Poindexter J. S.: *Strategies and Mechanisms for Field Research in Environmental Bioremediation,* American Academy of Microbiology, Washington, 1994.

Mueller, J. G.: *Model Studies of Bioremediation Strategies and Monitoring Technologies in Proceedings, Environmental Biotechnology,* EC-US Task Force on Biotechnology Research, Belgium, 1995.

National Research Council (NRC): *Alternatives for Ground Water Cleanup,* National Academy Press, Washington, 1994.

Old, R. W. and S. B. Primrose: *Principles of Gene Manipulation, An Introduction to Genetic Engineering,* University of California Press, California, 1980.

Olson, Steve: *An Industry Comes of Age,* National Academy Press, Washington, 1986.

Organisation for Economic Cooperation and Development (OECD): *Biotechnology for a Clean Environment, Prevention, Detection, Remediation,* OECD, Paris, 1994.

Ouellette, Robert P. and Paul N.: *Cheremisinoff, Essentials of Biotechnology, Lancaster,* Technomic Publishing Company, Pennsylvania, 1985.

Panem, Sandra: *Biotechnology, Implications for Public Policy,* The Brookings Institution, Washington, 1985.

Powell, W. and P. Brantley: *Competitive Cooperation in Biotechnology, Learning Through Networks,* Harvard Business School Press, Boston, 1992.

Prentis, Steve: *Biotechnology, A New Industrial Revolution,* George Braziller, New York, 1984.

President's Council on Competitiveness: *Report on National Biotechnology Policy,* The Council, Washington, 1991.

Rissler, Jane, and Margaret Mellon: *Perils Amidst the Promise,*

Ecological Risks of Transgenic Crops in a Global Market, Union of Concerned Scientists, Cambridge, 1993.

Saaty, R. P. and Booth S. R.: *Cost Effectiveness of a Remediation Technology Field Tested at the Savannah River Integrated Demonstration Site,* Los Alamos National Laboratory, New Mexico, 1994.

Saunders, Venetia A. and Jon R. Saunders: *Microbial Genetics Applied to Biotechnology, Principles & Techniques of Gene Transfer & Manipulation,* Crook Helm Ltd., London, 1987.

Saxenian, A.: *Regional Advantage,* Harvard University Press, Boston, 1994.

Sayler, G. S., Fox, R. and Blackburn, J. W.: *Environmental Biotechnology for Waste Treatment,* Plenum Press, New York, 1991.

Sayler, G. S.: *Biodegradation Process Analysis, Molecular Application in Simulations and Environmental Verification in Proceedings, Environmental Biotechnology,* EC-US Task Force on Biotechnology Research, Belgium, 1995.

Sayler, Gary S.: *Environmental Biotechnology for Waste Treatment,* Plenum Press, New York, 1991.

Skladany, G. J. and Metting, F. B.: *Bioremediation of Contaminated Soil in Soil Microbial Ecology,* Marcel Dekker, New York, 1992.

Smith, George P.: *The New Biology, Law, Ethics and Biotechnology,* Plenum Press, New York, 1989.

Sussman, M.: *The Release of Genetically-engineered Micro-organisms,* Harcourt Brace Jovanovich, New York, 1988.

Thomas, J. M. and Ward, C. H.: *In Situ Biorestoration of Organic Contaminants in the Subsurface,* Environmental Science and Technology, 1989.

Thompson, P: *Animal Welfare and Animal Rights, Center for Biotechnology Polilcy and Ethics,* Texas A&M University, Texas, 1992.

Thompson, Paul B., Robert J. Matthews: *Ethics, Public Policy and Agriculture*, MacMillan, London, 1994.

Tunlid, A. and White, D. C.: *Biochemical Analysis of Biomass, Community Structure, Nutritional Status, and Metabolic Activity of Microbial Communities in Soil in Soil Biochemistry*, Marcel Dekker, New York, 1992.

Wijesinghe, A. M., Knapp, R. B. and Carman, L. M.: *Preliminary Feasibility and Cost Analysis of the In-Situ Microbial Filter Concept*, Lawrence Livermore National Laboratory, Livermore, California, 1992.

Index

C

G

H

I

K

L

M

N

O

P

□□□